# Statistics in Dentistry

D1627975

# Statistics in Dentistry

## John S. Bulman, PhD, BDS

*Department of Community Dental Health, Institute of Dental Surgery,
Eastman Dental Hospital, 256 Gray's Inn Road, London WC1X 8LD*

## John F. Osborn, PhD, BSc

*Department of Epidemiology and Population Science, London School of
Hygiene and Tropical Medicine, Keppel Street, London WC1E 7HT*

*( now at Istituto di Igiene, Città
Universitaria, P Le Aldo Moro 5, 00185 Rome, Italy )*

Published by the British Dental Association
64 Wimpole Street, London W1M 8AL

Papers from the
*British Dental Journal*
January 21 to June 10, 1989

First printing 1989
Reprinted 1997


ISBN 0 904588 22 X


Typeset by E T Heron (Print) Ltd,
Silver End, Witham, Essex


Printed in Great Britain by
Biddles Ltd, Guildford and King's Lynn

# Contents

# Preface

Initially, this book was designed to assist general dental practitioners who read the British Dental Journal to understand the statistical content of published research papers. At the same time, it was hoped that some help would be given to researchers wishing to know not only what analysis they should apply to their own data, but also why they were applying it. Many of the chapters appeared in a recent series of articles in the British Dental Journal, but from correspondence and suggestions which followed, it was decided to include additional material, such as the chapter on survival analysis and a set of problems and answers to test the reader's understanding.

It was anticipated that the book would be of assistance to those who, in the past, have sought professional help with their statistical problems, but have been frustrated by a failure to understand the professional jargon. The authors also hoped, through this book, to reduce the number of papers submitted to dental journals, which, on refereeing, have had to be rejected or returned for correction owing to statistical errors and omissions. Finally, since many professional journals continue to publish papers containing unacceptable statistical analyses, it is hoped that this book will help all readers to distinguish the gold from the dross.

J. S. Bulman
J. F. Osborn

# Acknowledgements

Our thanks must first go to Margaret Seward, Editor of the British Dental Journal, who first proposed this idea to us, talked us into it, and then, over the next few years, listened patiently and cynically to our many excuses for non-delivery, until at last, in desperation, we delivered. Our warmest gratitude must also go to Dr Susan Silver, Assistant Editor, whose efficiency, understanding and unfailing cheerfulness made the chore of proof-reading almost enjoyable.

# 1

# Descriptive statistics

The importance of a knowledge of statistical methods in relation to most aspects of dentistry is discussed. Definitions of terms used in descriptive statistical techniques are provided, together with a review of the techniques themselves.

Among the many problems which beset the Scientific Advisers of a publication such as the *British Dental Journal,* one in particular recurs with regularity. If the journal is to maintain its reputation, only scientific papers of the highest calibre can be published; this means that, inevitably, papers may refer to techniques and methods of analysis unfamiliar to the majority of its readers. As a result, that majority readership may, not unreasonably, experience a feeling of irritation when faced with information which is of relevance to their work but which, because of the jargon used, borders on the unintelligible. To this irritation may be added a dark suspicion that the authors may be trying the not unknown gambit of blinding with science.

Probably one of the most frequent generators of such irritation today is the use of statistical techniques, with their reference to the mysteries of standard errors, *t*-tests, or regression analysis, often with a freedom which is in inverse proportion to the amount of explanation offered by the authors. Unfortunately, the suspicion that there may indeed be, as Disraeli observed, 'lies, damn lies, and statistics' is often well-founded. In an investigation reported in the *British Medical Journal* in 1977,[1] a team which included a qualified statistician analysed 62 reports which had appeared as papers and originals in 13 consecutive issues of that well-respected journal and which included statistical analyses. Of these, no less than 32 were found to have statistical errors, of which 18 were rated as 'serious'. Five of the reports made claims which were 'unsupportable on re-examination of the data'.

These and similar reports can easily give rise to such glib phrases as 'You can prove anything with statistics'. However, the errors referred to in the *British Medical Journal* paper were not errors of statistical technique, but rather errors in the selection and use of statistical techniques, and in the interpretation placed on the results obtained. To condemn an enormously valuable research tool just because it is frequently misapplied bears much the same sort of logic as executing a perfectly reliable messenger who is unlucky enough to bring bad news. We have to accept the fact that, for better or worse, statistical analysis and its interpretation are increasingly becoming an integral and essential part of the process of providing dental care throughout the world. Remuneration, evaluation of both new and old techniques and medicaments, manpower utilisation, performance appraisal, and the determination of treatment needs and demands, especially in the light of rapidly changing patterns of oral disease, all depend ultimately and completely on a proper appreciation of the implications of raw statistical data. The proper interpretation of such data requires a knowledge of the methods of statistical analysis, a knowledge which few dentists possess.

Unfortunately, not all those who undertake statistical analysis in print (or elsewhere), have received any formal statistical training either. Dental surgeons would be hesitant, to say the least, before permitting a first-year dental student with no preliminary training to perform a Class II cavity preparation on one of their patients; yet those same dental surgeons permit statistical assumptions which can affect their whole livelihood to be made by equally untrained individuals, without protest. Thus, it becomes very important that readers of dental journals or reports, who are probably dental surgeons first and statisticians a long way second, should at least be able to distinguish between the skilled and intelligent user of statistical techniques and the researcher, administrator, or armchair theorist who, as Andrew Lang once aptly put it, use statistics 'as a drunken man uses lampposts—for support rather than illumination'.[2]

Initially, it is necessary to ask the question 'Why is statistical analysis necessary?'. Lord Rutherford once remarked 'If your experiment needs statistics, you ought to have done a better experiment'.[3] Indeed, such analysis finds its place far more commonly in the life sciences than in the fields in which Lord Rutherford moved. If a chemist completes a series of repeated titrations and finds that his results differ to any important degree, he must assume that his apparatus or his methods were faulty, or his chemicals were impure or improperly mixed, or else that his skill was inadequate. However, if a dental surgeon gives a series of patients exactly the same dose of a local anaesthetic, using exactly the same technique, he would be very surprised if all his patients responded in exactly the same way, and he certainly would not assume, if the response varied, that this was necessarily due to a faulty technique or to an impure batch of anaesthetic. He would be far more likely to assume, and with good reason, that he was dealing with a case of natural variation, the same natural variation which decrees that one mature person may be five feet eight inches tall and another six feet one inch. Yet, if the response of a group of people to a given drug is naturally variable, how can we reasonably compare two alternative drugs with a similar action, to determine which is the most effective? This is but one of many situations where statistical techniques provide the only way of finding a reliable solution to the problem.

The purpose of this book is therefore not primarily to teach statistics, but rather to attempt to explain the logic behind statistical analysis, so that non-mathematical readers may at least come to terms with papers that rely on such analysis. Some mathematical formulae will inevitably have to be introduced from time to time, but these will never venture beyond the boundaries of elementary algebra. By the end, it is hoped that readers will be familiar with the terms and techniques most commonly used in statistical analysis, and will be in a position to judge for themselves whether they can understand and accept a statistical report, whether they need to refer to an expert for a second opinion, or whether they should reject it out of hand as inadequate.

## Variables

Before embarking on a review of the statistical techniques themselves, it is important to consider first the raw material upon which they will be used —numerical information, or, to use the more familiar term, data. Incidentally, data is the plural of datum, and it is therefore incorrect to say 'the data is . . .' when you mean 'the data are . . .'. Statistical data consist of observations made on independent units or individuals. The inclusion of the word 'independent' here is important. For example, a statistical analysis of dental disease using teeth as the units would be invalid, because teeth are not independent units. They are usually clustered together in groups of 32 or less, within very varying oral environments, and the basic independent investigatory unit in these circumstances has to be the mouth, or the individual, and not the tooth (unless, of course, only one tooth per mouth is being investigated). Teeth, in other words, have to be considered statistically as subunits of the unit from which they come— the mouth.

Often, a number of different observations are made on the same individual. For example, in a survey of dental health in children, a child's age, sex, social status and educational level may all be noted in addition to the DMF score. Any one of these observations is known as a *variable*. Variables which take a numerical value (such as age, weight, or DMF score) are called *quantitative variables*, while those that take a non-numerical value or define a characteristic (such as sex or social class) are known as *qualitative variables*. It is important to note here that arbitrarily assigning a numerical value to a qualitative variable does not automatically convert it into a quantitative variable. Thus, if the qualitative variable 'sex' is arbitrarily assigned a score of 1 to all males and 2 to all females, it does not thereby become a quantitative variable for the reason that, in this case, 2 can in no way have twice the magnitude of 1. This seems almost too obvious to be worth mentioning, and yet it is still possible to see, in print, reports in which qualitative periodontal diagnoses have been arbitrarily assigned numerical values or grades and then analysed as if they were quantitative data. This point will be returned to later in the book.

Just as variables may be qualitative or quantitative, so quantitative variables may be either *discrete* or *continuous*. A discrete variable can only take certain fixed numerical values within the range of observations. Thus, a child's DMF score can only take a whole number value, because it is dealing with numbers of whole teeth associated with a mouth, and therefore an individual child's DMF score cannot, for instance, be 5·2356. On the other hand, it would be perfectly possible for the width of a central incisor to be, say, 4·28 mm, because here the measurement can take any value along a continuous scale, and width, as a measure of length, is a continuous variable. Indeed, the only thing which limits the precision of measurement here (ie the number of decimal places that can legitimately be quoted in the reported width) is the sensitivity of the measuring device. Obviously, if tooth width is being measured with an ordinary millimetre rule, then widths quoted to two decimal places are not realistic, since you cannot measure hundredths of a millimetre with a millimetre rule. Yet, with disturbing regularity, it is possible to see statistics reported in scientific papers to an accuracy which is obviously quite beyond the capacity of the method of measurement used.

The distinction between discrete and continuous variables can be illustrated by considering age. In theory, age is a continuous variable, because it can obviously take any value along a continuous time scale. In practice, however, the tendency is to refer to age as 'age last birthday', so a person remains 30 until the day of their 31st birthday. Thus, reported age is almost always a discrete variable, and, unlike almost any other continuous variable, it is systematically 'rounded down' rather than rounded off to the nearest whole number. In other words, if someone is 16 years and 11 months we still refer to them as being 16 rather than 17.

## Descriptive statistics

Suppose that we have collected a large number of observations of incisor width (a continuous, quantitative variable), and now wish to look a little more closely at the pattern they present. One obvious way would be to arrange the observations into some sort of order, starting with the smallest width and ending with the largest. However, if we had a hundred or more measurements, this would make a very long list, and in any case some of the widths might be identical. The usual practice is therefore to prepare a *frequency distribution*. This is a statement of the number of observations (ie measurements) occurring at different values of incisor width. With continuous data, the range must be broken up into suitable intervals. The number of intervals used depends on convention (if any), and convenience. Usually, anything between 7 and 20 intervals are used, but this would also depend on sample size. A frequency distribution of central incisor widths is shown in Table I. Notice that the intervals used here increase by a constant amount of half a millimetre. These intervals are known as the *class intervals*. The dash (–) after any given width signifies 'up to, but not including' the next highest value listed. Thus, 18 incisors were observed within the interval 4·0 mm and up to but not including 4·5 mm wide, whereas 46 incisors were at least 4·5 mm wide but not 5·0 mm or wider.

The column heading 'Rel. frequency (%)' in Table I lists the frequencies for any given interval converted into a percentage. The first interval contains 18 observations, which represent 4% of the total number of observations (449). The fact that the relative frequency column does not

**Table I** Frequency and relative frequency distribution for a continuous variable (incisor width)

| Width mm | Frequency | Rel. frequency (%) |
|---|---|---|
| 4·0– | 18 | 4·0 |
| 4·5– | 46 | 10·2 |
| 5·0– | 58 | 12·9 |
| 5·5– | 110 | 24·5 |
| 6·0– | 120 | 26·7 |
| 6·5– | 60 | 13·4 |
| 7·0– | 32 | 7·1 |
| 7·5–8·0 | 5 | 1·1 |
| Total | 449 | 99·9 |

add up to exactly 100% is due to 'rounding off' the calculations of the interval percentages to one decimal point. Relative frequencies are useful when comparing data obtained in two or more different groups with different numbers of observations. Tables II and III show frequency distributions for (a) discrete quantitative data (dmf scores), and (b) a qualitative variable (colour of tooth stains). It should be noted here that dmf score is quantitative because it is a count of the number of times a particular qualitative categorisation (decayed, missing, or filled deciduous teeth) has been observed in an individual. Thus, the frequency distribution and relative frequency distribution provide a useful tabular presentation of the pattern of variability that has been observed. This simple first step in data summary is easily extended to provide a diagrammatic representation of the distribution. Such a diagram is called a *histogram*.

Histograms are easy to draw and interpret, provided the widths of the class intervals are all equal. Some problems can arise, and misleading diagrams may lead to incorrect conclusions, if care is not taken with

**Table II** Frequency and relative frequency distribution for a discrete variable (dmf score, children aged 5 years)

| dmf | Frequency | Rel. frequency |
|-----|-----------|----------------|
| 0 | 125 | 39·1 |
| 1–2 | 58 | 18·1 |
| 3–4 | 58 | 18·1 |
| 5–6 | 38 | 11·9 |
| 7–8 | 22 | 6·9 |
| 9–10 | 13 | 4·1 |
| 11–12 | 6 | 1·9 |
| Total | 320 | 100·1 |

**Table III** Frequency and relative frequency distribution for a qualitative variable (tooth stain colour)

| Stain | Frequency | Rel. frequency |
|-------|-----------|----------------|
| None | 225 | 82·1 |
| Orange/brown | 32 | 11·7 |
| Green | 12 | 4·4 |
| Black | 5 | 1·8 |
| Total | 274 | 100·0 |

histograms with unequal intervals. For example, the data in Table IV show the age (last birthday) at loss of last tooth in a sample of edentulous patients. Notice that the widths of the class intervals are not all equal. If a diagram were drawn with the heights of the blocks equal to the observed relative frequency, the block corresponding to age group 35–44 years would be the tallest, and it would be tempting to conclude that the most likely age at which the last tooth is lost would fall within this interval. However, a closer inspection of the table reveals that 20% lost their last tooth in the five years 30–34 compared with 29% in the ten years 35–44. It is hardly surprising that a larger percentage falls into a wide age range than falls into a narrow one! It is therefore necessary to make allowances for the width of the class interval. In general, the easiest way of doing this is to calculate the percentage occurring in a single unit of time for each interval, as shown in the final column of Table IV.

If these percentages per unit time are now used to construct a histogram, the result will be as shown in figure 1. Notice that the widths of the blocks in the histogram still correspond to the widths of the class intervals, and notice too that in a correctly drawn histogram, the area of the blocks is proportional to the (relative) frequency, and that the total area of the blocks is the total frequency or, in the case of relative frequencies, 100%. Sometimes the histogram is simplified by joining the midpoints of the tops of the blocks by straight lines. This is called an ogive or *frequency polygon*, which may be a useful way of comparing several relative frequency distributions in a single diagram.

**Table IV** Relative frequency distribution of age (last birthday) at loss of last tooth in a sample of edentulous men

| Age | Class interval | % | % per 1-year interval |
|-----|---------------|-----|----------------------|
| 11–15 | 5 years | 0·5 | $0·5 \times 1/5 = 0·10$ |
| 16–19 | 4 years | 3·5 | $3·5 \times 1/4 = 0·875$ |
| 20–24 | 5 years | 10·5 | $10·5 \times 1/5 = 2·10$ |
| 25–29 | 5 years | 17·5 | $17·5 \times 1/5 = 3·50$ |
| 30–34 | 5 years | 20·0 | $20·0 \times 1/5 = 4·00$ |
| 35–44 | 10 years | 29·0 | $29 \times 1/10 = 2·90$ |
| 45–54 | 10 years | 14·0 | $14 \times 1/10 = 1·40$ |
| 55–74 | 20 years | 5·0 | $5 \times 1/20 = 0·25$ |

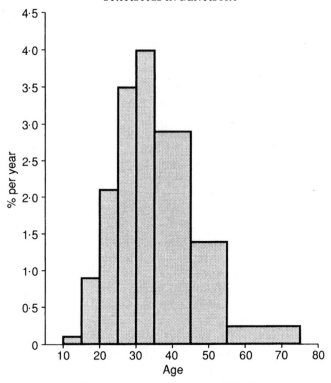

**Fig. 1** Histogram of data from Table IV.

In the next chapter, further aspects of the position and dispersion of data in a frequency distribution will be discussed, and the terms mean, median, mode and standard deviation will be introduced.

### References

1 Gore S M, Jones I G, Rytter E C. Misuse of statistical methods: critical assessment of articles in *BMJ* from January to March 1976. *Br Med J* 1977; **1**: 85–87.
2 Mackay A L. *The harvest of a quiet eye. A selection of scientific quotations.* p91. Bristol: Institute of Physics, 1977.
3 Bailey N J T. *The mathematical approach to biology and medicine.* New York: Wiley, 1967.

# 2

# Simple summary calculations

Methods of measuring the general magnitude or central tendency of sets of data are described, including the mean, median and mode. Similarly, methods of determining the degree of dispersion of the observations about the mean, including range, variance, and standard deviation are presented, together with worked practical examples of their use and importance. The use of algebraic notation in statistics is described.

The frequency distribution described in the first chapter provides a useful basis for the derivation of other ways of describing data. For example, now that the largest and smallest observations in a set of data are readily apparent, it becomes possible to calculate the *range* of the observations. The range, defined statistically as the difference between these two extreme values, is normally used for data description rather than analysis. For example, if in a sample of children the lowest DMF score is 2 and the highest 15, the range of scores will be from 2 to 15 or, by statistical definition, $15-2 = 13$ teeth. This range from 2 to 15 is a measure of both position (ie magnitude) and dispersion (ie spread), but it is not a very efficient statistic, as it tells us nothing about the size or distribution of the observations between these two extremes.

A more important descriptive statistic, which provides an indication of the general magnitude of the observations within the distribution is some kind of average value. There are several sorts of average, including the arithmetic mean, the median, and the mode. Notice that the word 'average' is used in this context in a generic sense, to include all measures of overall magnitude.

For example, given the gross annual income of each of ten dental surgeons, the *arithmetic mean* is the sum of all ten incomes, divided by ten. The *median*, in contrast, is determined by arranging all the ten annual incomes in ascending or descending order of magnitude and then selecting the middle value in the ordered series (or the mean of the middle two values if, as in this case, there is an even number of observations). The *mode* is the most 'popular' observation (income in our example), that is, the one that occurs most frequently. The mode is probably the least useful, since many distributions, and especially those based on a small number of observations, may have more than one mode, simply as a result of chance. For example, the following distribution is unimodal—that is, it has only one mode (4):

$$2,2,3,3,3,4,4,4,4,5,5,5,6,6,$$

whereas the next is bimodal, since the most 'popular' observations are 3 and 5, each of which occurs four times:

$$2,2,3,3,3,3,4,4,4,5,5,5,5,6,6,$$

yet both distributions follow essentially the same pattern.

In contrast, the mean is the most useful statistic of the three, if only because it makes the most use of the data available, since the magnitude of each and every observation contributes to the calculated mean value.

Before going any further, it is worth taking a few moments to think about what these 'average' values actually tell us about a distribution. They can be very misleading to anyone who does not understand their limitations. A good example of this is provided by the dentists' incomes referred to earlier. Suppose the mean income is calculated as £30 000 per annum. Such a figure could be obtained from widely varying income distributions. It could, for example, be the result of roughly half the dentists earning between £29 000 and £30 000 per annum and the rest earning between £30 000 and £31 000. Alternatively, it could mean that most dentists earned £27 000 or less, but a small minority earned £37 000 or more. Both distributions (and indeed many others as well) could give a mean income of £30 000. Again, suppose the mean DMF of a group of 500 12-year-old children is reported to be 3·45. This could indicate that each child has a DMF approximating more or less to 3·45; however, suppose, as is increasingly common these days, about 50% of the children were completely caries free. The mean DMF of those with caries would then be closer to 6·90! Reporting caries epidemiological data in terms of the mean DMF for the whole population, without specifying what percentage of the population is caries free, is to present a most misleading picture. Yet this is almost routine procedure in published dental epidemiological reports, dating back to the days when virtually no child was caries free. A more realistic approach today would be to report the percentage of subjects who are completely caries free, and then give the mean DMF for the remainder.

In some cases, the median may provide a more reliable indication of the true average. Thus, in the second example of dentists' incomes given above, the median would almost certainly be nearer to £27 000, which is a much better indicator of the majority income than the mean of £30 000. This is a good illustration of the fact that statistics, when misused or misunderstood, can be misleading. In the same way that a computer is frequently blamed for the faults of its programmer, so statistics are regularly blamed for the shortcomings of their interpreters.

Here is another little anomaly which has trapped many authors of dental statistical papers. Reference was made in the first article to the problem of age as a discrete variable. If ages are calculated as 'age last birthday', and if we are presented with an even distribution of schoolchildren aged 10 to 12 years, then their mean age will not be 11, as might be expected, but actually $11 \cdot 5$ (or 11 years 6 months). This is because we are effectively dealing with a continuous age range of 3 years rather than 2 years, 10–11, 11–12, and 12–13, because, by definition, the child remains aged 12 right up to their 13th birthday. On the other hand, if the ages had been exact, ie a continuous variable ranging from $10 \cdot 0$ years to $12 \cdot 0$ years, then the mean age would indeed have been $11 \cdot 0$ years.

At this point it is necessary to introduce some algebra, in order to describe some of the statistical symbols met with most frequently in scientific papers. Earlier it was noted that the arithmetic mean was calculated by adding all the observations of a variable together and dividing this sum by the number of observations. In algebraic terms this might be written as

$$\frac{x_1 + x_2 + x_3 \ldots + x_n}{n}$$

where $x_1$, $x_2$, $x_3$ are the values observed for the variable $x$ which we are interested in, and $n$ is the number of observations made. Thus, in the dentists' income example mentioned earlier, $x$ would represent the variable 'income', $x_1$, $x_2$, $x_3$ would represent the first three incomes recorded, and $n$ would be 10, the total number of incomes recorded.

However, even with only ten observations this is a rather clumsy formula, so instead of writing $x_1 + x_2 + x_3 \ldots x_n$ on the top line, statisticians use the 'shorthand' notation of $\Sigma x$, where $\Sigma$, the Greek capital letter sigma, means 'the sum of'. So $\Sigma x$ means 'the sum of all the values of $x$'. Notice that $\Sigma$ is an instruction, not a quantity; $\Sigma x$ does not mean $\Sigma$ multiplied by $x$. Finally, the symbol for the mean value of a sample of observations of the variable $x$ is usually written as $\bar{x}$, called '$x$ bar'). The formula for the arithmetic mean of a variable $x$ in a sample of size $n$ is therefore

$$\bar{x} = \frac{\Sigma x}{n}$$

Before leaving the subject of average or mean values, one final important point has to be made. Supposing in a school dental inspection you examine 60 boys and 40 girls. You find that all the children have at least one carious tooth and you calculate that the mean DMF of the boys is $5 \cdot 1$ and the mean DMF of the girls is $4 \cdot 4$. What, then, is the mean DMF of the whole group? In answering this question you might be tempted to say that the overall mean DMF is $(5 \cdot 1 + 4 \cdot 4)/2$ = $4 \cdot 75$. However, this would only be correct if there were an equal number of boys and girls. As it is, we have to allow for the fact that more boys (60) have contributed to their mean DMF of $5 \cdot 1$ than girls (40) have contributed to their mean DMF of $4 \cdot 4$. So it becomes necessary to *weight* each mean value according to the number of children who have contributed to it. This is done by first multiplying each mean value by the number of children contributing to it, then adding the resulting two values together, and finally dividing this sum by the total number of children:

$$\frac{(5 \cdot 1 \times 60) + (4 \cdot 4 \times 40)}{60 + 40} = 4 \cdot 82$$

The values 60 and 40 are often called 'weights' and $4 \cdot 82$ is the weighted mean of $5 \cdot 1$ and $4 \cdot 4$.

The arithmetic mean is a summary value of a series of quantitative observations. If the data are qualitative, the corresponding summary value is a *proportion*. Thus, if in a series of 30 cases of congenitally missing teeth there are 21 males and 9 females, the proportion of the sample who are male is 21/30 or $0 \cdot 7$. This may also be written as 70% (ie $0 \cdot 7 \times 100$). An algebraic formula for this calculation would therefore be $p = r/n$, where $p$ is the proportion, $r$ is the number of males and $n$ is the total sample size. In general, a proportion may be calculated whenever we have a sample of $n$ observations, where each observation can be classified into one of two possible categories which may be generalised as *success* and *failure*. In the above example we have $n = 30$; the characteristic 'male' is regarded (for this example only!) as being labelled 'success', and so the proportion is calculated as the number of successes, $r$, divided by the total sample size $n$. In a sample of $n$ independent observations of a qualitative variable where the possibility of either 'success' or 'failure' exists, and the chance of each individual being labelled 'success' is constant, the number of successes $r$ out of $n$ observations is said to be a *binomial* variable.

## Algebraic notation in statistics

We have so far met some simple algebraic formulae which used the following symbols: $x$, $\bar{x}$, $n$, $p$, $r$ (all of which may have subscripts), and the operator $\Sigma$. Unfortunately, there are no absolute rules about the choice of symbols used to denote variables, although almost universally $\Sigma$ means 'the sum of'. (Surprisingly, in the best of the many basic textbooks of medical statistics,[1] A. Bradford Hill uses the word 'sum' rather than $\Sigma$.) There are, however, conventions for the choice of symbols and the most important convention arises out of problems concerned with sampling. For example, there exists in this country a 'population' of dental surgeons, from which a 'sample' might be drawn in order to calculate the income data referred to above. It is important to understand that the population and the sample are two separate groups. The object of sampling is to use the information provided by the sample in order to give some estimate of the nature of the population. So the arithmetic mean derived from the sample might be used as an *estimate* of the arithmetic mean of the income of all dentists in the population. Any measure which describes a characteristic of a population (such as mean, median, proportion, or variance and standard deviation, which we will be coming to shortly) is known as a *population parameter*, whereas similar measures derived from samples are known as *statistics*, or sometimes as *sample estimates*. Thus, for each parameter in the population, a statistic calculated from the sample may be available as an estimate. In order to differentiate clearly between the two, algebraic notation usually makes use of Roman letters when describing sample statistics, and Greek letters when describing population parameters. A population mean is therefore identified as $\mu$, the lower case Greek letter 'mu', whereas a sample mean is designated, as already described, $\bar{x}$ ($x$ bar). A population proportion is denoted by the lower case Greek $\pi$ (nothing to do with circles or geometry in this context), while, as we have seen, the sample proportion is $p$. The Greek letter for the population parameter usually corresponds to the Roman letter used for the sample estimate. We deal with samples because obtaining data from a whole population is usually far too expensive in terms of time, manpower and money.

Incidentally, the term 'parameter' is frequently misused by non-statisticians to describe observations of a variable, or even the variable itself. Variables such as height, weight, age, Periodontal Index, gingival pocket depth, or dentist's income might erroneously be referred to as 'parameters'. Such misuse may provide a useful indication of who is statistically reliable and who is not!

## Summarising the dispersion or variability

When discussing the limitations of the mean, median and mode, attention was drawn to the fact that very diverse patterns of figures can produce identical mean values. Thus each of the following samples of $n = 5$ observations has a mean value of 5:

$$3, 4, 5, 6, 7 \quad \text{and}$$

$$2, 2, 2, 2, 17.$$

It therefore becomes necessary, when describing a set of data, to invoke yet another descriptive measure, one which provides an indication of the *variability* of the observations within each sample. One such measure is the range, but because it is based on only the two most extreme values (which are likely to be the least reliable) and ignores the others, it is clearly inefficient. An alternative approach would be to look at the divergence of each observation from the arithmetic mean. These divergences will be large if there is a lot of variability, but relatively small if the values cluster closely around the mean.

Consider a *population* of $N$ individuals, each of whom is associated with a value of the variable $x$. We have seen that the population mean $\mu$ is $\Sigma x / N$. Then the mean of the differences $x_1 - \mu$, $x_2 - \mu$, . . . $x_n - \mu$ might, on the face of it, be used to provide a measure of the mean variability of the $N$ values of $x$. However, this turns out to be impractical, as the positive differences (which arise from values of $x$ greater than the mean $\mu$) will exactly cancel out the negative difference. One way around this problem is to square the differences (since the square of a number, whether it is itself positive or negative, is always positive) and calculate the mean value of these squared differences. This quantity is called the *population variance* and it is denoted by the lower case Greek letter sigma ($\sigma$) squared.

$$\text{pop. var.}(x) \quad = \quad \sigma^2 \quad = \quad \frac{\Sigma(x - \mu)^2}{N}$$

In practice it is unlikely that every value of $x$ in the population will be known, so the population mean $\mu$ will therefore also be unknown and it will be impossible to calculate $\sigma^2$. If a random *sample* of $n$ individuals is taken from the population, their values of $x$ can provide an *estimate* of $\sigma^2$. Note, however, that the average value of $\Sigma(x-\bar{x})^2$ in the sample will tend to systematically underestimate the average value of $\Sigma(x-\mu)^2$ in the population. That is, there will tend to be less variability in a sample than in the population. This bias is overcome by calculating the *estimated variance*, denoted by the Roman letter $s^2$, as:

$$\text{est. var.}(x) = s^2 = \frac{\Sigma(x-\bar{x})^2}{n-1}$$

The divisor in the estimated variance, $n-1$ in this example, is called the *'degrees of freedom'* and may be thought of as the number of independent deviations from the sample mean. Thus, if we have $n$ observations of variable $x$, giving $n$ values of $x-\bar{x}$, all the information of these differences is contained in any $n-1$ of them. This is because any one difference can always be calculated if the others are known, since the total of the $n$ differences (as was pointed out earlier) must be zero.

A small example can be used to illustrate the points made. Consider the first sample of $n=5$ observations quoted above, where $\bar{x}$ also equals 5. If we simply add together the deviations of each value from the mean we end up with 0:

| $x$ | $(x-\bar{x})$ |
|---|---|
| 3 | $-2$ |
| 4 | $-1$ |
| 5 | 0 |
| 6 | 1 |
| 7 | 2 |
| $\Sigma x = 25$ | $\Sigma(x-\bar{x}) = 0$ |

However, if the value $(x-\bar{x})$ is squared, the sum must be positive, since the square of a negative number is always positive:

| $x$ | $x-\bar{x}$ | $(x-\bar{x})^2$ |
|---|---|---|
| 3 | $-2$ | 4 |
| 4 | $-1$ | 1 |
| 5 | 0 | 0 |
| 6 | 1 | 1 |
| 7 | 2 | 4 |
| Total | | $10 = \Sigma(x-\bar{x})^2$ |

Since $n=5$, the estimated variance ($s^2$) will be $10/4=2\cdot5$.

Notice that the units of variance are the square of the original units in which the variable $x$ was measured. So if $x$ is measured in metres, the estimated variance is expressed in square metres. In order to provide a measure of the variability in the units of the original observations, all that has to be done is to find the square root of the variance. This quantity is known as the *standard deviation*. In the population, this is denoted algebraically as $\sigma$, and the sample estimate of $\sigma$ is denoted by $s$. So the formula for the sample estimate of the standard deviation is:

$$s = \sqrt{\frac{\Sigma(x-\bar{x})^2}{n-1}}$$

Thus, in the second example given above, where $x$ took the values $2,2,2,2$ and $17$, it follows that the estimated variance of the observations is $45\cdot00$ and the standard deviation is $\sqrt{45\cdot00}$ which is $6\cdot71$. So although both of these samples have a mean of 5, the first, with values close to the mean, has a standard deviation of $1\cdot58$ (ie $\sqrt{2\cdot5}$), whereas the second sample, with values more widely dispersed about the mean, has a standard deviation of $6\cdot71$. The wider the dispersion of values of $x$ about the mean, the larger the standard deviation. Furthermore, in many of the distributions which are met with in the life sciences, a rough rule of thumb is that about 95% of all observations in such distributions will lie within the interval $\bar{x}\pm2s$, that is, between $\bar{x}-2s$ and $\bar{x}+2s$. This point will be returned to in the next of these articles.

A practical example which might be of relevance to dental practitioners concerns the usable shelf-life of a new filling material. Suppose material 'A' has a mean usable shelf-life of 25 weeks, whereas material 'B' has a mean usable shelf-life of 27 weeks. On the face of it, 'B' would appear to be the better bargain; however, if the standard deviation was 5 for 'B' but 1 for 'A', this would mean that the actual shelf-life for most samples of 'B' might be anything between 17 and 37 week, whereas most of the samples of 'A' would have a shelf-life varying only between 23 and 27 weeks. In these circumstances, 'A' at once becomes the more reliable product.

One further practical point remains. When calculating the variance or the standard deviation, it would be exceedingly tedious to have to calculate each individual difference between $x$ and $\bar{x}$, especially in distributions where the number of observations was large, and where $\bar{x}$ was unlikely to be a simple whole number. For this reason, a 'short-cut' formula can be used if the calculations are to be done by hand, although today a computer program or an electronic calculator is much more commonly used:

$$s^2 = \frac{\Sigma x^2 - [(\Sigma x)^2/n]}{n-1}$$

So, again, using the first example we have

| $x$ | $x^2$ |
|-----|-------|
| 3 | 9 |
| 4 | 16 |
| 5 | 25 |
| 6 | 36 |
| 7 | 49 |
| $\Sigma x = 25$ | $\Sigma x^2 = 135$ |

$(\Sigma x)^2/n = 25^2/5 = 125$

$\Sigma(x-\bar{x})^2 = \Sigma x^2 - (\Sigma x)^2/n$

$\phantom{\Sigma(x-\bar{x})^2} = 135 - 125 = 10$

$s^2 = 10/(5-1) = 2\cdot 5$

$s = \sqrt{2\cdot 5} = 1\cdot 58$

Although this 'short-cut' formula may look more complicated, it makes the calculation of the variance using a very simple form of electronic calculator much faster and easier. Most modern hand-held calculators have programs for this calculation built into them; this eliminates the tedium of calculating the standard deviation almost completely.

In chapter 3, some probability distributions which are of importance in statistics will be considered, and the statistical implications of sampling will be discussed.

## Practical example

In a laboratory test, 20 uniform bar samples of an alloy proposed for use in denture construction are subjected to increasing stress by mounting one end of the bar in a vice and hanging increasing weights from the free end. The weight (in kilograms) required to fracture the sample ($x$) is noted and the results are as follows:
(NB '$f$' represents the frequency of occurrence of each value of $x$)

| $x$ | 1 | 2 | 3 | 4 | 5 | 6 | $n$ |
|-----|---|---|---|---|---|---|-----|
| $f$ | 3 | 4 | 6 | 3 | 3 | 1 | 20 |

Calculate the mean, median, and modal weights required to fracture the samples. What is the estimated standard deviation of the weight distribution? Would you be inclined to use this material as a denture alloy?

## Reference

1 Hill A B. *A short textbook of medical statistics*. London: Hodder and Stoughton, 1977.

# 3

# Probability and sampling

The concept of probability is introduced, and the role of probability distributions in statistical theory is discussed, with particular reference to the Normal distribution and its characteristics. Sampling and sampling variation are described, together with sampling error, the standard error of the mean, and confidence intervals for determining the likely magnitude of a population mean.

## Probability

This topic lies at the very core of the subject of statistics. It has fascinated philosophers, mathematicians, gamblers and many others, for hundreds of years. At its simplest it is not difficult to understand, and a good idea of its statistical meaning can be gleaned from the everyday usage of the word. If we toss a coin, we know that the *probability* of it landing 'head' side up is one half, or 50%, since unless the coin lands on its edge there are only two alternative ways for it to fall—'heads' up or 'tails' up, and each of these is equally likely. However, it may be useful to define probability a little more strictly. The probability of an event (eg getting a head when a coin is tossed) is the proportion of times this event would occur in a long series of random trials (coin tosses).

Notice that a probability is a proportion. Its value must therefore lie between 0 and 1 (or 0% and 100%). If a probability is 0, the event never occurs. If it is 1, the event always occurs. Also, notice the words 'long series'. In a short series of say five tosses of a coin we may get three heads and two tails, but the probability of a head would not therefore be 3/5 = 0·6.

If we consider the probabilities of each and every possible outcome, the sum of these probabilities must be 1 (or 100%). In tossing a coin, the only possible outcomes are a head or a tail. The probability of a head is 0·5 and the probability of a tail is 0·5 and the sum of these is 1·0. Any breakdown of the total probability 1·0 is called a *probability distribution*. A more practical example might be the probability distribution of DMF scores in, say, 10-year-old boys. If we had a very large sample of such boys and recorded their DMF scores, then the proportion with no DMF teeth, the proportion with one DMF tooth, the proportion with two DMF teeth and so on, should be a good estimate of the probability distribution of the DMF scores in the population. In general, the relative frequency distribution obtained from a sample provides us with an estimate of the underlying population probability distribution.

We have considered here both a qualitative variable (heads or tails) and a discrete quantitative variable (DMF). What about a continuous variable? A slight problem can present itself if we consider probability distributions of truly continuous variables. For example, what is the probability that the height of a man selected from the population at random is exactly $59 \cdot 3842976$ inches? The answer is presumably zero. The proportion of men having that exact height in even a very large sample is almost certainly zero. This means that when considering a continuous variable, probability can only refer to intervals on the continuous scale. The probability that a man is between 60 and 62 inches tall is then just the proportion of men whose heights fall within this range.

There are a large number of theoretical probability distributions derived from mathematics, which are often used as models to describe data that are observed in practice. Among the most common of these theoretical distributions is the *Normal* distribution. This is by far the most important, because not only is it useful as a model for much observed data, but also because it provides the underlying theory for significance tests and sampling errors, which we will be considering in the following chapters.

## The Normal distribution

The Normal distribution can be defined by an algebraic formula which is too complicated to be of interest to us here. However, from this formula it is possible to determine that a normal distribution is defined by two quantities, its arithmetic mean ($\mu$) and its standard deviation ($\sigma$). Its general shape is rather like the cross-section of a bell, a wide bell if the standard deviation is relatively large, and a narrow bell if the standard deviation is relatively small. This is shown schematically in figure 1. Note that the distribution is symmetrical about the mean, and that the mean = the median = the mode.

For example, the probability distribution of heights of adult men closely resembles a Normal distribution, with the mean ($\mu$) equal to about 68 inches and standard deviation ($\sigma$) equal to about $2 \cdot 5$ inches. This will look

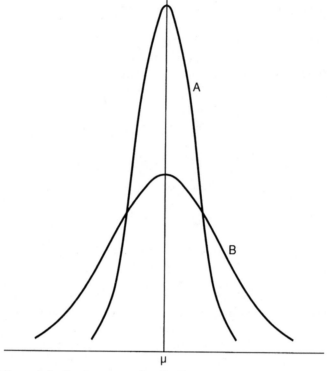

**Fig. 1** Normal distribution curves for distributions with (A) small and (B) large standard deviations.

something like figure 2. As expected, this shows that the heights of most adult men peak close to the mean of 68 inches, and the proportions of extremely short men (ie less than 60 inches) or extremely tall men (ie more than 76 inches) are relatively small. The distribution is symmetrical about the mean of 68 inches. Because of the underlying mathematical definition of the Normal distribution, some of its properties can be listed. For example:

(a) The proportion of men shorter than the mean height of 68 inches is equal to the proportion taller than the mean height. That is, the mode and the median are equal to the arithmetic mean, and the distribution is symmetrical about this central value.

(b) The proportion of men within the interval $\mu \pm \sigma$, that is, 65·5 inches to 70·5 inches is 0·68 or 68%. Thus, 16% of men are shorter than 65·5 inches and 16% are taller than 70·5 inches.

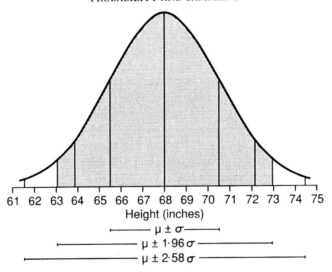

**Fig. 2** Distribution curve of heights of adult men.

(c) The proportion of men within $\mu \pm 1 \cdot 96\sigma$, that is $63 \cdot 1$ inches to $72 \cdot 9$ inches, is 95%, so that $2 \cdot 5\%$ are shorter than $63 \cdot 1$ inches and $2 \cdot 5\%$ are taller than $72 \cdot 9$ inches.

(d) Finally, the proportion of men within $\mu \pm 2 \cdot 58\sigma$, or $61 \cdot 5$ to $74 \cdot 4$ inches is 99%, so that $0 \cdot 5\%$ are less than $61 \cdot 5$ inches tall and $0 \cdot 5\%$ are more than $74 \cdot 4$ inches tall.

Other, similar, intervals can be worked out using tables of the Normal distribution, which can be found in nearly all statistical texts. In summary, the most important intervals are: (a) $\mu \pm 1 \cdot 96\sigma$ includes 95% of observations, leaving $2 \cdot 5\%$ in each 'tail' (this is the basis for the rough rule given in the previous chapter), and (b) $\mu \pm 2 \cdot 58\sigma$ includes 99% of observations, leaving $0 \cdot 5\%$ in each 'tail'.

The constants $(1 \cdot 96, 2 \cdot 58)$ found in the formulae given above are known by the somewhat strange name of *standard normal deviates,* and they will be considered in more detail shortly.

## Sampling
Chapter 2 made reference to sampling, mainly in connection with the use of algebraic symbols which differentiated between parameters referring to population, and statistics referring to samples. It is now necessary to consider the relationship between populations and samples in more detail.

*Sampling* may be defined as the investigation of part of a population, in order to provide information which can then be generalised to cover the whole population. More often than not in public health work, sampling may be the only way to obtain information about a population, because the true extent of the population is unknown, or, even if it were known, access to the whole population is impossible. Other reasons for working with samples rather than populations include the need to reduce labour and hence cost. It should, however, be appreciated that such (often considerable) savings in time, manpower and money are only achieved at the expense of replacing absolute and comprehensive data concerning the population with an approximation. Thus, if it were possible to examine every 12-year-old child living in a Health District, it could be reported with certainty that, for instance, the mean number of missing teeth per child was $2 \cdot 6$. If, on the other hand, it were only possible to examine a sample of 10% of the children, then it might well be found that the mean number of missing teeth in this 10% sample was $2 \cdot 2$. Obviously, in these circumstances, it would be wrong to say 'because the sample mean for tooth loss is $2 \cdot 2$, I can therefore assume that the true population mean is also $2 \cdot 2$'. What is needed is some way of determining the reliability or precision of our sample mean when used as an *estimate* of the true population mean. How this may be done will now be considered.

First of all, let us be clear on what we mean by 'population' and 'sample'. A population is the total group with the characteristic in which we are interested. It might be composed of people, animals, houses, or hospital records, and must be divisible into a set of non-overlapping parts, such that the individual measurements for the parts in total constitute the population. These parts are called *units,* and the list of all units is called the *sampling frame.* Units in our 12-year-old child population would be individual 12-year-old children, but in other situations might also be household addresses, or schools.

A sample should be drawn at random from the population. A non-random, or 'purposive' sample, where apparently 'typical' units may be selected for study, may well provide accurate data, but there exists no way of determining just how accurate. A basic definition of random sampling is sampling in which each and every unit in the sampling frame has an independent and equal chance of being included in the sample. There are many ways of achieving this, most of them involving the use of random numbers, and it should be noted that true randomisation is, paradoxically, a most precise operation. It is not, for example, sufficient to take a list of names and then prick it 'randomly' with a pin to determine who will be in the sample. Such a method would tend to favour those names printed near the centre of the sheet and neglect those printed at the corners and edges. Numerous sampling strategies exist, and you will often see reference to 'simple random sampling', 'stratified sampling', 'cluster sampling' or 'multi-stage sampling'. Regrettably, space does not permit further discussion on this topic, although errors in sampling methods are nearly as common in research reports as errors in statistical methods.

There are two important reasons why a sample should be random. First, a random sample will avoid bias, which is a systematic tendency to overestimate or underestimate the population parameter. Secondly, with a random sample, statistical techniques can be used to make probability statements about the population parameter. This is the basis of significance tests and confidence intervals. For the moment, suffice it to say that one of the main assumptions made when initiating statistical analysis is that the data to be analysed have been obtained from a random sample of the population. Although examples of statistical analysis being carried out on non-random or 'biased' samples may be met with in dental and medical literature, the fact remains that the amount of useful information which can be derived from such analysis may be very limited. This means that you should be very cautious about believing statistical analysis of data derived, for instance, from a postal questionnaire which has only achieved a 70% response. The missing 30% may well be more than enough to create sufficient bias in the 'sample' to ensure that it is no longer representative of the population from which it is drawn.

The most common characteristics of a population which we may wish to estimate by taking a sample are (a) the mean value of some measurement, and (b) the proportion of the population with some characteristic. As already mentioned, such sample means or sample proportions are most unlikely to equal exactly the true population means or proportions. We will have to deal with what is known as *sampling error.*

Here you may well be wondering what the difference is between sampling error and any other sort of error. Sampling errors are due to the fact that we have observed only a selection of units drawn from the whole population. It therefore follows that sampling errors tend to diminish and become less important as the sample size increases. If 200 individuals in a population are examined, the results will be more reliable than if only 20 are examined. In general, sampling error tends towards zero as the sample size tends towards the population size.

By way of contrast, non-sampling errors do not necessarily get smaller as sample size increases. Examples of such errors could occur if, for instance, in an adult dental survey a high proportion of elderly ladies refused to cooperate, or if in a child survey based on schools, only state schools as opposed to private schools were available for examinations. Another common source of non-sampling error is where one or more investigators use different diagnostic criteria in their examinations (*see* chapter 9). Such events would introduce greater or lesser systematic errors, or biases into the sample, and so reduce the value of the findings, unless, as can happen occasionally, it is possible to measure reasonably accurately the extent of the bias. For example, if all the uncooperative elderly ladies in the adult survey were known to be edentulous, allowance for this could be made when calculating the mean DMF.

## Sampling variation

So far we have discussed drawing one sample of size $n$ from a population and using it to determine $\bar{x}$, the sample mean. We have also noted that $\bar{x}$ will rarely be exactly equal to $\mu$, the population mean. Now suppose for the moment that we were to take several different samples, all of size $n$, and all from the same population. We would end up with several different values of $\bar{x}$, and if these were similar, we would infer that the sampling error is probably small; in other words, any one of the sample means is likely to be close to the population mean $\mu$. On the other hand, if the sample means differed widely, we would conclude that the sampling error is likely to be large, because any particular sample could provide an estimate far from the population mean $\mu$. We can make two intuitive judgements about such sampling errors:

(a) The sampling error will get smaller as the sample size ($n$) gets larger; ie big samples are more reliable than small samples.

(b) The sampling error depends on the variability of the observations. If the variability of the observations is small (ie if the variance and standard deviation are small), we would expect the sampling error to be small. If it were large, the sampling error may well also be large.

It is worth pursuing this idea a little further. Consider a population, each member of which is associated with the value of a variable $x$. (If this algebraic approach worries you, think of the variable $x$ as, say, a person's height.) If we wish to estimate the true mean $\mu$, of $x$ (height), we might do so by taking a sample size $n$ and calculating $\mu = \Sigma x/n$. If this routine were repeated for very many different samples of size $n$, we would end up with a whole frequency distribution of different values of $\bar{x}$. If we used these data to plot a histogram of $\bar{x}$, instead of $x$ as described in the first chapter, it could be shown mathematically that:

(a) The distribution of $\bar{x}$ tends to be Normal, even if the parent distribution of $x$ is markedly non-Normal. The distribution of means comes closer to the Normal distribution as $n$ increases in size.

(b) The mean of the distribution of $\bar{x}$ is the same as the mean of the distribution of $x$ (which is $\mu$).

(c) The variance of the distribution of $\bar{x}$ can be shown to equal $\sigma^2/n$, where $\sigma^2$ is the variance of the distribution of $x$. Thus, as the sample size $n$ increases, so the variance of the distribution of $\bar{x}$ will decrease, as we would expect.

(d) The standard deviation of the distribution of $\bar{x}$ is once again the square root of the variance (see chapter 2), and this is often called the *standard error of the mean*. It is usually written as $SE(\bar{x})$.

$$SE(\bar{x}) = \sigma/\sqrt{n}$$

Since $\bar{x}$ tends to be Normally distributed, we are able to apply all the properties of a Normal distribution to distributions of the sample mean. Thus, we can expect 95% of all sample means to lie within the range $\mu \pm 1\cdot96(SE(\bar{x}))$. Or, put another way, in only about 1 in 20 cases will the sample mean be either greater or less than $1\cdot96$ standard errors on either side of the population mean, that is, lie outside the range

$$\mu \pm 1\cdot96(\sigma/\sqrt{n})$$

This argument is vital to an understanding of all that follows, so a little recapitulation may be helpful.

Suppose we have a large 'population' of individuals in late middle-age, and direct measurement on each individual shows that the mean number of missing teeth ($\mu$) is $5\cdot0$, and the distribution of the number of missing teeth per person is skewed to the right. The curve might look something like figure 3. We can calculate the variance ($\sigma^2$) and the standard deviation ($\sigma$) for the distribution (as described in the previous chapter), and these might be, for instance, $4\cdot0$ and $2\cdot0$, respectively.

Now suppose that we were able to take a series of samples of, perhaps, nine individuals ($n=9$) from this population, and calculate the mean number of missing teeth for each sample of nine ($\bar{x}$). We would end up with a whole series of different values of $\bar{x}$, one for each sample we took. These might take such values as $4\cdot2$, $5\cdot6$, $6\cdot4$, $5\cdot1$, and so on. Once we had enough of them, we could plot another frequency distribution curve, but this time not of the individual tooth loss values as we did above, but rather

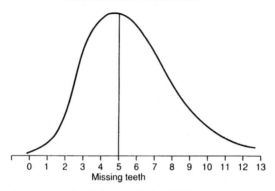

**Fig. 3** Distribution curve of missing teeth in middle-age.

of the *mean* tooth loss values for each of our samples of nine individuals. We have already noted that in such a distribution, (a) the distribution will be Normal, (b) the mean will be $\mu$ (in this case 5), (c) the standard deviation of this sampling distribution will be the standard error, which may be calculated as $\sigma/\sqrt{n}$ (in this case $\sigma/\sqrt{9} = 2/3 = 0.67$). Thus, the resulting distribution would look something like figure 4.

From figure 3 we can see that, very roughly (especially as the distribution deviates somewhat from true Normality), 95% of the measurements of tooth loss in the population will lie within the range $\mu \pm 1.96\sigma = 5 \pm 1.96 \times 2 = 1$ to 9, while more precisely (since they are more Normally distributed), 95% of repeated sample means of size $n = 9$ will lie within the range $\mu \pm 1.96\sigma/\sqrt{n} = 5 \pm 1.96 \times 2/3 = 3.7$ to $6.3$.

Obviously, in practice we would never go to all the trouble of taking large numbers of samples and calculating $\bar{x}$ for each sample; we are likely to have to make do with just one sample. That one sample is nevertheless one observation from the theoretical distribution of sample means as described above. From this we know that there is a 95% chance that the mean value of any one sample drawn from a known population will lie within the range

$$\mu \pm 1.96 \times \text{SE}(\bar{x})$$

or, put another way, that there is a 95% chance (ie $0.95$ probability) that the error in using $\bar{x}$ as an estimate of $\mu$, will not be greater than $1.96 \times \text{SE}(\bar{x})$.

If this statement is now turned around and looked at from another angle, it follows that, given a single sample mean $\bar{x}$, there is a 95% probability that the range

$$\bar{x} \pm 1.96 \times \text{SE}(\bar{x})$$

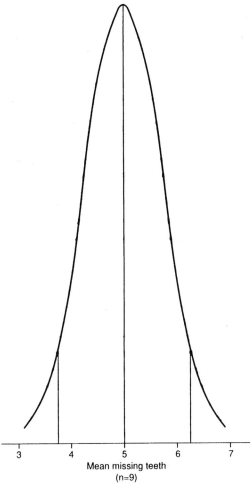

3  4  5  6  7

Mean missing teeth
(n=9)

**Fig. 4** Distribution curve of sample means of missing teeth.

will include the true population mean $\mu$, and this is because the maximum likely error is $1 \cdot 96 \times SE(\bar{x})$.

The interval, $\bar{x} \pm 1 \cdot 96 \times SE(\bar{x})$, is known as the *95% confidence interval* for $\mu$ and is a measure of the precision of the sample mean $\bar{x}$ as an estimate of the population mean $\mu$. If a wider interval is wanted, with a higher probability of including the true value of $\mu$, then the 99% confidence interval may be calculated. The formula is exactly the same, except that the constant $1 \cdot 96$ is replaced with $2 \cdot 58$. These constants, you will recall, have already been referred to as standard normal deviates, and the values for any given probability level (ie 95%, 99%) may be found listed in tables in most statistical textbooks or published in their own right. In our example, we can therefore be 99% certain that $\bar{x} \pm 2 \cdot 58 \times SE(\bar{x})$ will include the population mean $\mu$.

In the next chapter we will be looking at some practical applications of these so far mainly theoretical considerations.

**Practical example**

The standard deviation of the fluoride level in samples of water from all sources within an administrative region is known from long experience to be 0·03 ppm. The mean fluoride level of a random sample of 12 water sources is found to be 0·40 ppm. Within what limits would you expect the mean fluoride level for the whole region to lie?

# 4

# Simple tests of statistical significance

Simple methods of determining whether the difference between a sample mean and a given standard value is likely to be due only to sampling error, or to be real and/or of practical importance, are described. The use of the null hypothesis and of standard normal deviates is discussed, and examples are provided to illustrate the points made.

So far, we have been concerned with ways of describing and reporting data, together with a series of theoretical concepts, the practical applications of which may not have been readily apparent to readers. It is hoped that the importance of these concepts will now become clear, as we deal with the more practical aspects of analytical statistics.

The introduction of the *standard error of the mean* in the previous chapter makes it possible to predict the likely magnitude of a true population mean, given the mean of one sample drawn from that population. Moreover, it enables us to make informed decisions on whether a given sample is likely to have originated from a given population, or whether two independent samples are likely to have been drawn from the same population. A basic example of the first of these two latter problems now follows.

## Single sample, σ known

Suppose that in a given health district, the mean number of restorations placed per NHS practitioner per month is known to be 214, with a standard deviation ($\sigma$) of ten. In one group practice of five dental surgeons, however, the mean number of restorations placed per practitioner per month is 231. In these circumstances, is there any evidence to suggest that the dentists in this practice differ systematically from their colleagues in the same district in terms of the number of restorations placed, or could the difference between the population mean ($\mu$) of 214 and the sample mean ($\bar{x}$) of 231 be attributable merely to sampling variation?

In order to answer this question, let us for the moment assume that there is no real difference in the number of restorations placed in the group practice, compared to the rest of the district, and that the practice can therefore be regarded as a random sample of five, drawn from all the dentists in the district. This assumption is known in statistical terms as setting up a *null hypothesis*. From the previous chapter we know that in these circumstances, means of samples of size 5 are distributed Normally about the population mean ($\mu = 214$), and that the standard deviation of this distribution of sample means, ie the standard error of the mean, is

$$\sigma/\sqrt{n} = 10/\sqrt{5} = 4 \cdot 47$$

We also know that in a Normal distribution, sample means outside the range $\mu \pm 1 \cdot 96\text{SE}(\bar{x})$, that is $214 \pm 1 \cdot 96 \times 4 \cdot 47$, or $205 \cdot 2$ to $222 \cdot 8$, would occur in only 5% of samples of this size, ie with probability $P = 0 \cdot 05$. Our sample mean (231) lies outside this range, so what can we conclude? There are two possibilities:

(1) Our null hypothesis (that the difference is the result of sampling variation) is correct and the sample mean is so large simply because we were unlucky in drawing a sample whose mean happened to lie at the upper extreme of the sample distribution;

(2) Our null hypothesis is wrong.

We cannot be absolutely sure which of these two alternatives is correct, but since the probability of a sample mean as extreme as 231, given that statement (1) is correct, is less than 5% (remember that only 5% of all sample means will be larger than 223 or smaller than 205), we would be inclined to accept alternative (2). Thus, we *reject* the null hypothesis and conclude that, on the basis of this sample, it appears that, on average, the dentists in the group practice are placing more restorations than their colleagues elsewhere in the same district.

This type of argument is called a *test of significance*. In general terms, what we have done is to ask whether a sample with mean $\bar{x}$ could reasonably have come from a population with mean $\mu$ and standard deviation $\sigma$ and, in this case, we have decided that it could not.

If (as here) $\bar{x}$ does not lie within the interval $\mu \pm 1 \cdot 96\sigma/\sqrt{n}$, or, equivalently (doing a little mathematical juggling), if

$$\frac{\bar{x} - \mu}{\sigma / \sqrt{n}}$$

is numerically greater than $1 \cdot 96$, we say that the difference between $\bar{x}$ and $\mu$ is statistically *significant* at the 5% level; this is written in statistical shorthand as $P < 0 \cdot 05$ (ie the probability of observing a sample mean as extreme as 231 if the null hypothesis is true is less than $0 \cdot 05$, or 5%). P stands for 'probability', and < means 'is less than'. Similarly, > means 'is greater than'.

If we wanted to be even more certain that our conclusion in this example was the correct one, we could make use of the fact that we would expect only 1% of means of samples of this size to occur outside the range $214 \pm 2 \cdot 58 \times 4 \cdot 47 = 202 \cdot 5$ to $225 \cdot 5$. The sample mean we have calculated, 231, is even outside this expanded range, so we can still reject the null hypothesis, but now the corresponding value of $P$ is $< 0 \cdot 01$.

In other words, if $(\bar{x} - \mu)/(\sigma/\sqrt{n})$ is greater than $2 \cdot 58$, then the difference is significant at the 1% level; this is recorded in statistical shorthand as $P < 0 \cdot 01$. These critical *standard normal deviates* (SNDs), $1 \cdot 96$ and $2 \cdot 58$, associated with the 5% and 1% levels of probability, respectively, are chosen for this testing purpose quite arbitrarily. You could, for example, just as easily base your significance test on the 10% level of probability, using the SND constant of $1 \cdot 64$, so that if

$$\frac{\bar{x} - \mu}{\sigma / \sqrt{n}}$$

is greater than $1 \cdot 64$, you could reject the null hypothesis with the knowledge that there is only a 10% chance that you are wrong in so doing. However, a mere 5% chance of being wrong (odds of 20 to 1 against) is a much more convincing argument than a 10% chance of being wrong (odds of only 10 to 1 against). Therefore, by custom, the 5% level of probability has become accepted as the arbitrary dividing line between, on the one hand, rejecting a null hypothesis and, on the other, concluding that there is insufficient evidence safely to reject it.

It should be noted that one cannot use this procedure to prove that a null hypothesis is correct. If in the example quoted the calculated SND was substantially less than $1 \cdot 96$, we could not therefore assume that we had proved that our sample of five dentists was not placing significantly more fillings than their colleagues; we have merely been unable to produce adequate statistical evidence to suggest that they were.

If the calculated SND lies between $1 \cdot 96$ and $2 \cdot 58$, the level of significance is between 1% and 5% and we write $0 \cdot 01 < P < 0 \cdot 05$.

It is important to note that in this example, even though in statistical terms the group practitioners were placing significantly more restorations than their colleagues, there could be many practical reasons for this, and it would be wrong to jump to too hasty conclusions. We have not, for example, proved that the dentists in the group practice are guilty of overprescribing, since they might, for example, be treating a higher than average number of patients with a high level of caries susceptibility. This particular result is therefore only the first step in a far more complex investigatory process, if the true reason for the difference in treatment patterns is to be determined. By itself, it merely indicates that the pattern of restoration placement within this group practice probably differs from that of the rest of the district. This is a very important point, as it is not uncommon to find authors coming to the most far-reaching conclusions on cause and effect, on the strength of a positive test of significance which, in itself, provides little if any evidence to support them.

It is also important to note that if the standard deviation had been, say, 21 instead of 10, we would not have been able to reject our null hypothesis at the 5% level, since the standard error of the mean would then have been 9·39, giving an SND of 1·81; this is markedly less than the critical value of 1·96 needed for a significant result.

The term 'significant' in this context has the special meaning of 'not easily explained by sampling error', rather than the more general meaning of 'important' or 'consequential'. This can also lead to misunderstanding; one group of patients may have significantly more carious teeth than another in statistical terms, but if the difference is only of the order of a mean of 0·5 carious teeth, then this is probably of no real clinical significance or importance. For this reason, whenever presenting the result of a significance test in a written report, it is almost invariably advisable to report also the 95% confidence interval for the difference between the means (or proportions) under investigation, since this gives a measure of the actual range within which we would expect the true population difference to lie; this makes it easier to assess whether the statistical significance is important in real terms. It also makes it easier for non-statisticians to understand what is going on.

This underlines the point that, at the very least, it is discourteous for authors to present statistical results in terms of statistical jargon (ie Result sig. $P < 0·05$) without explaining what is meant in non-statistical terms; this is especially true in journals such as the *BDJ*, where the majority of the readership should not be expected to be qualified statisticians!

In the worked example above, the 95% confidence interval for the true difference between the population and sample means will be

$$(\bar{x} - \mu) \pm 1 \cdot 96 \times SE(\bar{x})$$

$$= (231 - 214) \pm 1 \cdot 96 \times 4 \cdot 47$$

$$= 17 \pm 8 \cdot 76$$

$$= 8 \cdot 2 \text{ to } 25 \cdot 8$$

In other words, this five-practitioner group would be expected to place, on average, not less than eight and not more than 26 fillings per practitioner per month more than the average for the district. Had the standard error of the mean been smaller, we would have been able to be more precise in this estimate; had it been larger, we would have been less precise. Notice also from the formula $SE(\bar{x}) = \sigma/\sqrt{n}$, the influence that the sample size will have in this context.

We could, if we so wished, also calculate the 99% confidence interval for the true difference, by substituting $2 \cdot 58$ for $1 \cdot 96$ in the above formula.


## Single sample σ not known

The previous example was somewhat artificial, in the sense that it gave the population standard deviation, $(\sigma)$, which in real life is almost never known. However, if you don't known $\sigma$, how can you calculate $SE(\bar{x})$? The answer is to use the best available estimate of $\sigma$, which is s, the *sample standard deviation*. This is not too unreasonable, since whether the observations come from the whole population or a sample drawn from it, they will still show the same pattern of spread about the mean, especially if the sample is a large one.

The formula for the *estimated standard error of the mean* is therefore $s/\sqrt{n}$ instead of $\sigma/\sqrt{n}$. Common sense decrees that s will provide a more reliable estimate of $\sigma$ if the sample is large than if it is small. To compensate for this increasing unreliability of s as an estimate of $\sigma$ in small samples, the SNDs, $1 \cdot 96$ and $2 \cdot 58$, which we have used up to now in significance testing, have to be replaced with other critical values, derived from what is known as the *t-distribution*. Unlike the SNDs, the *t*-values depend on the sample size or, more precisely, in statistical terminology, the degrees of freedom in $s^2$ as an estimate of $\sigma^2$. The critical values of $t$ become larger as the samples get smaller for any given level of significance, thus compensating for the increasingly poor precision of s as an estimate of $\sigma$ as sample sizes get smaller:

| 5% level | Crit. Val. of SND | Crit. Val. of $t_{(n-1)}$ |
|---|---|---|
| $n = 100$ | 1·96 | 1·98 |
| $n = 10$ | 1·96 | 2·26 |

| 1% level | | |
|---|---|---|
| $n = 100$ | 2·58 | 2·63 |
| $n = 10$ | 2·58 | 3·25 |

$$\text{While SND} = \frac{\bar{x} - \mu}{\sigma/\sqrt{n}} \qquad t_{(n-1)} = \frac{\bar{x} - \mu}{s/\sqrt{n}}$$

95% CI for $\mu$ is $\bar{x} \pm 1\cdot96 \text{ SE}(\bar{x})$ or $\bar{x} \pm t_{(0\cdot05,\, n-1)} \text{ Est.SE}(\bar{x})$

99% CI for $\mu$ is $\bar{x} \pm 2\cdot58 \text{ SE}(\bar{x})$ or $\bar{x} \pm t_{(0\cdot01,\, n-1)} \text{ Est. SE}(\bar{x})$

The addition of the subscript $(n-1)$ to $t$ indicates that when you look up a $t$-value in the relevant statistical tables, you do not relate it to the actual sample size you are dealing with, but rather to the degrees of freedom. Thus, if you do a one-sample $t$-test involving a sample of size 20, the calculated $t$-value is reported as being 'on 19 df', and has to be compared with the critical $t$-value for 19 degrees of freedom for the required level of significance in the tables (*see* Appendix 1). For instance, if, in the previous example, we were not given $\sigma$, but were told instead that the sample standard deviation (s) was 8·67:

$$\text{then Est.SE}(\bar{x}) = 8\cdot67/\sqrt{5} = 3\cdot88$$
$$t = (231 - 214)/3\cdot88 = 4\cdot38 \text{ on 4 df}$$

From statistical tables we find that

$$t_{(0\cdot05,4)} = 2\cdot78 \qquad t_{(0\cdot01,4)} = 4\cdot60$$

Since our calculated $t$-value of 4·38 lies between the 5% critical value of 2·78 and the 1% critical value of 4·60, we would therefore still reject the null hypothesis, but in this case at the probability level $0\cdot01 < P < 0\cdot05$. There is less than a 5% but more than a 1% probability of such an extreme value occurring by chance.

Note that the sign (plus or minus) of the calculated $t$-value is irrelevant to the test; the magnitude of the value is all that matters. The sign is useful only insofar as it identifies which of the mean values being compared is larger.

The formula for calculating the 95% confidence interval will now be

$$(\bar{x} - \mu) \pm t_{(0.05,4)} \times \text{Est.SE}(\bar{x})$$
$$= (231 - 214) \pm 2.78 \times 3.88$$
$$= 17 \pm 10.79$$
$$= 6.2 \text{ to } 27.8$$

There are two important assumptions which should be borne in mind when using the $t$-test:
(1) It is assumed that the sample(s) are taken from populations whose values are Normally distributed (although the test is robust enough to remain reliable even when there is some departure from exact Normality).
(2) It is assumed that variance(s) within the population(s) from which the samples are taken are the same.
The second assumption is the more important of the two. It is often found that in distributions which diverge from Normality, the standard deviation increases proportionally with the magnitude of the observations, ie large values are more variable than small values. This is particularly true of DMF data in dental epidemiological investigations, which do not follow a true Normal distribution but are 'skew to the right'; that is, there are more subjects with high DMF counts than would be found in a true Normal distribution. For this reason, DMF data have unstable variances, since the variance increases as the mean increases, and should not therefore be subjected to $t$-testing as they stand (although they almost invariably are in published reports!). There is a way around the problem, using logarithmic transformation of the raw DMF data, and this will be described in the next chapter.

### Matched pairs
So far, we have been considering problems involving a single sample mean, and relating it to a single population mean. However, more usually we will be comparing two samples with each other, rather than a sample with a population. A dental example might be to compare the efficiency of two different coagulant dressings following the surgical extraction of impacted wisdom teeth. Although it is possible to make the comparison by using two separate samples of patients, it is also possible to treat patients who are undergoing bilateral extractions with both coagulant dressings, one on each side. The comparison is therefore made within the same patient, with the

patient acting as his or her own control. This is an ideal form of 'matching'. In another example, comparing the cariostatic effect of one dentifrice with another, children within the population to be tested might be paired off, so that each pair was as similar as possible in terms of age, sex, baseline DMF, teeth present, social status, and any other factors likely to affect the incidence of new carious lesions. One child in each pair would then be placed in the 'study' group and receive the 'test' dentifrice, while the other would go into the 'control' group and receive the 'standard' dentifrice. At the end of the trial, the number of new carious surfaces in each child in the 'study' group ($a$) is subtracted from the number of new carious surfaces in their corresponding partner in the 'control' group ($b$), and the significance test carried out on these individual differences ($d = a - b$). The null hypothesis states that, on average, the number of new lesions in children in both groups will be equal, which, if true, indicates that in the population, the mean difference will be zero (ie $\mu = 0$)

If $a - b = d$ for each sample matched pair, then $\bar{d}$ will be the mean of all the values of $d$ (ie the equivalent of $\bar{x}$), and the $t$-test formula will be

$$t = (\bar{d} - 0)/\text{ Est.SE}(\bar{d})$$

The main advantage of matching subjects in a clinical trial like this is that the effect of the treatment or therapy is observed by making the comparison within each matched pair. In this way, the effect of disturbing variables such as age, sex, etc, is minimised. The disadvantage lies in the difficulty of arranging true matches in the pairings, and this increases dramatically as more factors have to be included in the matching process. In the above example, it would probably be almost impossible to obtain a large number of ideally matching pairs of children (ie same sex, age, DMF, social status, oral cleanliness level and so on).

In chapter 5, methods of comparing independent samples will be described, together with the logarithmic transformation of data and significance tests on proportions.

### Practical example

It is known, as a result of testing over many years, that the mean functional lifespan of an established dental operating lamp under normal working conditions is 356 hours. A new lamp has recently come on to the market, costing about 5% more, and a dental practitioner has tested ten of them. He finds that the mean functional lifespan of these ten lamps is 380 hours, with an estimated standard deviation ($s$) of 30·3 hours. Is it worth his while investing in the new lamps or should he stick with the old?

# 5

# Comparison of two independent means and two proportions

The comparison of two means obtained from either small or large samples and the comparison of two proportions is described, together with the use of logarithmic transformation of quantitative data in order to stabilise variance.

## Comparison of two sample means

So far in this review of tests of significance, we have compared a single sample mean with a population mean, or two matched or paired samples where the difference between the pairs can be treated as a single sample for analysis purposes. The next logical step in this process is to consider the comparison of two sample means when no such pairing is possible. An example might be a comparison of the oral health state of samples of individuals drawn from two different communities. Before considering this type of problem in detail, it is regrettably necessary to review some more statistical theory, concerning the mean and variance of the difference between two independent variables.

Suppose we have two distributions with means $\mu_1$ and $\mu_2$, and variances $\sigma_1^2$ and $\sigma_2^2$. If we then take samples of size $n_1$ and $n_2$ from the two distributions, we can calculate $(\bar{x}_1 - \bar{x}_2)$ for each pair of samples. If this were repeated often enough, we could then plot a distribution of these differences, in much the same way as we plotted, in theory, a distribution of sample means in chapter 3.

What could then be said about this distribution? It can be shown mathematically that:

(1) the mean of the distribution of $(\bar{x}_1 - \bar{x}_2)$ is $(\mu_1 - \mu_2)$;

(2) the variance of the differences, $\text{var}(\bar{x}_1 - \bar{x}_2)$, is $\text{var}(\bar{x}_1) + \text{var}(\bar{x}_2)$: ie the *sum* of the individual variances (and not the difference between them as you might reasonably expect at first sight)

$$\text{var}(\bar{x}_1 - \bar{x}_2) = \sigma_1^2/n_1 + \sigma_2^2/n_2$$

and the $\text{SE}(\bar{x}_1 - \bar{x}_2)$ will be the square root of this.

**Comparing sample means. (1) Large samples**

In the previous chapter we discussed significance tests which posed the problem: 'Is it likely that a sample of $n$ observations with mean $\bar{x}$ could have been taken from a population with mean $\mu$ and standard deviation $\sigma$?' This was answered by calculating

$$\text{SND} = \frac{\bar{x} - \mu}{\text{SE}(\bar{x})}$$

Now that two sample means are being compared, the null hypothesis to be tested is that 'the samples were taken from populations whose means $\mu_1$ and $\mu_2$ are equal', ie $\mu_1 = \mu_2$ or $\mu_1 - \mu_2 = 0$. In other words, the hypothesis proposes that the observed difference $\bar{x}_1 - \bar{x}_2$ is merely due to sampling error. To test this hypothesis, we again calculate the SND, but this time $\bar{x}$ is replaced by $\bar{x}_1 - \bar{x}_2$ and $\mu$ is replaced by $\mu_1 - \mu_2$ (which under the terms of our null hypothesis equals 0).

$$
\begin{aligned}
\text{SND} &= \frac{(\bar{x}_1 - \bar{x}_2) - 0}{\text{SE}(\bar{x}_1 - \bar{x}_2)} \\
&= \frac{\bar{x}_1 - \bar{x}_2}{\sqrt{(\sigma_1^2/n_1) + (\sigma_2^2/n_2)}} \\
&= \frac{\bar{x}_1 - \bar{x}_2}{\sqrt{\text{SE}^2(\bar{x}_1) + \text{SE}^2(\bar{x}_2)}}
\end{aligned}
$$

and the SND is interpreted as before.

A 95% confidence interval for the true difference, $\mu_1 - \mu_2$ is

$$(\bar{x}_1 - \bar{x}_2) \pm 1 \cdot 96\text{SE}(\bar{x}_1 - \bar{x}_2)$$

**Comparing sample means. (2) Small samples
($\sigma$ unknown)**

Here again we have to face the problem that, if we want to use the $t$-test, we must assume that:

(1) the populations are Normally distributed;
(2) the variances in the two populations are equal.

The first of these assumptions is again not usually important, but the second puts a constraint on the null hypothesis.

In general terms, the null hypothesis we are dealing with here would take the form: 'The two samples under review were drawn from Normal populations, whose means $\mu_1$ and $\mu_2$ are equal and whose variances $\sigma_1^2$ and $\sigma_2^2$ are also equal'. In other words, the two samples are taken from identical populations.

Under the terms of the null hypothesis, it then follows that

$$\sigma_1^2 = \sigma_2^2 = \sigma^2 \text{ (for convenience)}$$

Thus, our two sample variances, $s_1^2$ and $s_2^2$ are two separate estimates of the same quantity $\sigma^2$. Since it is rather illogical to have two separate estimates of the same quantity, it is not unreasonable to combine $s_1^2$ and $s_2^2$ to produce $s^2$, the single best estimate of $\sigma^2$.

$$\text{Since } s_1^2 = \frac{\Sigma(x_1 - \bar{x}_1)^2}{n_1 - 1} \qquad \text{and } s_2^2 = \frac{\Sigma(x_2 - \bar{x}_2)^2}{n_2 - 1}$$

it follows that the weighted average of $s_1^2$ and $s_2^2$, using $n_1 - 1$ and $n_2 - 1$, respectively, as weights is

$$s^2 = \frac{(n_1 - 1)s_1^2 + (n_2 - 1)s_2^2}{(n_1 - 1) + (n_2 - 1)}$$
$$= \frac{\Sigma(x_1 - \bar{x}_1)^2 + \Sigma(x_2 - \bar{x}_2)^2}{n_1 + n_2 - 2}$$

The estimated standard error of the difference between $\bar{x}_1$ and $\bar{x}_2$ will then be

$$\text{Est. SE}(\bar{x}_1 - \bar{x}_2) = \sqrt{(s^2/n_1) + (s^2/n_2)} = s\sqrt{(1/n_1) + (1/n_2)}$$

and this formula is used when calculating $t$:

$t = (\bar{x}_1 - \bar{x}_2)/\text{Est. SE}(\bar{x}_1 - \bar{x}_2)$ with $n_1 + n_2 - 2$ degrees of freedom.

A 95% confidence interval for the true difference between the population means $(\mu_1 - \mu_2)$ may be calculated as before, using the critical 5% $t$-value with $n_1 + n_2 - 2$ degrees of freedom.

A warning was given in the previous chapter, that $t$-tests should not be carried out on raw DMF data, because these were not Normally distributed and so had unstable variances. It was then suggested that this problem could be solved by carrying out a logarithmic transformation on such data before subjecting them to statistical analysis. This technique will now be described.

## The logarithmic transformation

The log transformation has four common uses, of which three are relevant here and one will come up in a later discussion:

(1) It enables comparisons to be made by ratio (or percentage) rather than by absolute difference. For example, in comparing the cariostatic effect of two toothpastes, A and B, we could look at the absolute difference between the mean DMF(S) increments, $\bar{x}_A - \bar{x}_B$, in two groups of children using A or B over a period of years. But would this really be an appropriate comparison or would not a *ratio* comparison of the form: 'on average, children using A had 30% more lesions than children using B' be more informative? Your first introduction to logarithms was probably in solving multiplication and division problems, when, to solve A/B, you calculated logA − logB and took the antilog of the answer, while A × B was the antilog of logA + logB, hence the ratio effect.

(2) The log transformation will often stabilise variance. The *t*-test of two independent sample means assumes that the variances $\sigma_1^2$ and $\sigma_2^2$ in the populations from which the samples were taken, are equal. This assumption is usually invalid in practice and the *t*-test may be misleading, because large values usually have greater standard deviations than small values. If the standard deviation is proportional to the magnitude of the value, ie the variance is not constant, the log transformation will stabilise the variance (as if by statistical magic!).

(3) Distribution of quantitative measurements is usually non-Normal. In practice, most distributions observed are skew to the right, ie they have a longer 'tail' on the right-hand side of the curve than on the left. This may not be very important for the *t*-test because the test is *robust* (that is, it will not usually let you down because the distribution is slightly skew). On the other hand, distributions which are skew to the right will approximate more closely to a Normal distribution if the log transformation is used.

## Application

The logarithmic transformation of some observation $x$ is defined by the formula $y = \log x$. There are two commonly used types of logarithm: logs to base 10 and logs to base e. Logs to base e are also called natural or Naperian logarithms. The symbol used for logarithms to base 10 is $\log_{10}$, while logarithms to base e are written either as $\log_e$ or ln. If you see 'log' written without a subscript, it may mean either base 10 or base e.

The antilogarithm of $y$ is $10^y$ if using logs to base 10, and $e^y$ if using logs to base e. All these functions are found on nearly all modern calculating machines. The letter e is used to denote a mathematical constant equal to

2·718281828 (approximately!), and some algebraic formulae can be written more tidily if $\log_e$ is used rather than $\log_{10}$. For our purposes, however, it usually does not matter which logarithm is chosen. Note that the log transformation can only be used for positive, non-zero values of $x$.

## Examples
### Two independent samples
One effect of the log transformation is to make constant proportional differences appear as absolute differences in the logarithm. For example, in an experiment we may have drug doses $x$:

| $x =$ | 2 | 20 | 200 | 2000 |
|---|---|---|---|---|
| $\log_{10}x =$ | 0·30 | 1·30 | 2·30 | 3·30 |
| $\log_e x =$ | 0·69 | 3·00 | 5·30 | 7·60 |

While successive doses are multiplied by a constant, 10, the log doses increase absolutely by 1 for $\log_{10}$ and 2·30 (apart from rounding error) for $\log_e$.

Ratio comparisons of dose are equivalent to absolute comparisons of log dose. Thus, for example, a comparison of two sample means can be made by taking the difference $\bar{x}_1 - \bar{x}_2$, and confidence intervals and/or significance tests may then be determined as previously described. If, however, the raw data are transformed using $\log_{10}$ or $\log_e$, and the mean logs compared absolutely, this difference between mean logs is the logarithm of a *ratio*. Significance tests and confidence intervals based on logarithms provide tests and confidence intervals of ratios.

The following data show the abrasiveness of two brush-on denture cleaners, A and B, in a standardised experiment. The response is denture weight loss in mg.

A: 10·2 11·0 9·6 9·8 9·9 10·5 11·2 9·5 10·1 11·8
B: 9·6 8·5 9·0 9·8 10·7 9·0 9·5 9·9

Applying the two-sample $t$-test (you may wish to check the calculations) we get:

$$\bar{x}_A = 10·36 \qquad \bar{x}_B = 9·50$$
$$n_A = 10 \qquad n_B = 8$$
$$s^2_A = 0·5716 \qquad s^2_B = 0·4571$$

pooled variance $s^2 = 0·5215 \qquad s = 0·7221$
Est. $SE(\bar{x}_A - \bar{x}_B) = 0·7221\sqrt{1/10 + 1/8}$
$$= 0·3425$$

To test the hypothesis that the denture cleaners are equally abrasive, we have:

$$t = (10 \cdot 36 - 9 \cdot 50)/0 \cdot 3425 = 0 \cdot 86/0 \cdot 3425$$
$$= 2 \cdot 51 \text{ with } (10 - 1) + (8 - 1) = 16 \text{ degrees of freedom}$$

which is greater than the critical value of $t$ with 16 df at the $P = 0 \cdot 05$ level (which is $2 \cdot 12$), so the hypothesis is rejected, $P < 0 \cdot 05$.
A 95% confidence interval for the true mean difference is

$$0 \cdot 86 \pm 2 \cdot 12 \times 0 \cdot 3425 \text{ which is } 0 \cdot 13 \text{ to } 1 \cdot 59 \text{ mg.}$$

If a ratio comparison of the abrasiveness of the denture cleaners were required, it would be necessary to take logs of the raw data. The following are $y = \log_e x$ of the observed values of weight loss ($\log_{10}$ could also be used and would give exactly the same result for $t$ and the confidence interval)

A: $2 \cdot 32 \ 2 \cdot 40 \ 2 \cdot 26 \ 2 \cdot 28 \ 2 \cdot 29 \ 2 \cdot 35 \ 2 \cdot 42 \ 2 \cdot 25 \ 2 \cdot 31 \ 2 \cdot 47$
B: $2 \cdot 26 \ 2 \cdot 14 \ 2 \cdot 20 \ 2 \cdot 28 \ 2 \cdot 37 \ 2 \cdot 20 \ 2 \cdot 25 \ 2 \cdot 29$

$$\bar{y}_A = 2 \cdot 336 \qquad \qquad \bar{y}_B = 2 \cdot 249$$
$$n_A = 10 \qquad \qquad n_B = 8$$
$$s^2_A = 0 \cdot 005121 \qquad \qquad s^2_B = 0 \cdot 005015$$
$$\text{pooled variance } s^2 = 0 \cdot 005075, \ s = 0 \cdot 07124$$
$$\text{Est. } SE(\bar{y}_A - \bar{y}_B) = 0 \cdot 07124 \sqrt{1/10 + 1/8}$$
$$= 0 \cdot 03379$$
$$\text{Difference } \bar{y}_A - \bar{y}_B = 0 \cdot 08653$$

To test the hypothesis that the ratio of the abrasiveness of A to that of B is one, we test the equivalent hypothesis, that the difference between mean logs is zero.

$$t = (2 \cdot 336 - 2 \cdot 249)/0 \cdot 03379 = 2 \cdot 56 \text{ with } 16 \text{ df}$$

which is interpreted in the same way as for the untransformed data. Since $t$ is greater than the critical value of $t$ with 16 df at the $P = 0 \cdot 05$ level, the difference between the mean logs is significantly different to zero, ie the ratio differs significantly from one.
A 95% confidence interval for the mean difference between the logs is

$$0 \cdot 08653 \pm 2 \cdot 12 \times 0 \cdot 03379 = 0 \cdot 01487 \text{ to } 0 \cdot 15819.$$

So far, the analysis is very similar to that given for the untransformed data, except that the analysis of logs does not give answers that are immediately interpretable. If such answers are required, the remedy is to use antilogs. Thus, the best estimate of the true ratio is antilog $(\bar{y}_A - \bar{y}_B)$ $= e^{0 \cdot 08653} = 1 \cdot 090$. Thus, on average, A is about 9% more abrasive than B. A 95% confidence interval for the true ratio is given by the antilogs of the limits for the difference between the mean logs, that is, antilog $0 \cdot 01487$ to antilog $0 \cdot 15819 = 1 \cdot 015$ to $1 \cdot 171$. A is therefore likely to be not less than $1 \cdot 5\%$ and not more than 17% more abrasive than B.

Notice that in the untransformed data we had

$$s^2_A = 0 \cdot 5716, \qquad s^2_B = 0 \cdot 4571$$

so that the two standard deviations are

$$s_A = 0 \cdot 7560 \text{ and } s_B = 0 \cdot 6761, \text{ with means}$$

$$\bar{x}_A = 10 \cdot 36 \text{ and } \bar{x}_B = 9 \cdot 50$$

As we discussed earlier, here is an example of a distribution where as the mean increases, so does the variance. If it can be shown that the standard deviation is proportional to the mean, that is if $0 \cdot 7560/10 \cdot 36$ is approximately equal to $0 \cdot 6761/9 \cdot 50$ (ie if $0 \cdot 073$ is approximately equal to $0 \cdot 071$, which it clearly is), then the log transformation should tend to make the variances in the two groups equal. This is the case here, since with the $\log_e$ transformation we have $s_A = 0 \cdot 005121$, which is approximately equal to $s_B = 0 \cdot 005015$, and so we can legitimately use the $t$-test.

*Matched pairs*
When a 'matched pair' procedure is used (as described at the end of the previous chapter), it may sometimes again be advantageous to make use of a log transformation of the data. In order to determine whether such a transformation would be advisable, the best approach is to draw a scatter diagram of the difference between the two responses (vertical axis) against the mean response (or, equivalently, the total of the two responses) on the horizontal axis. If the points tend to be evenly scattered about a horizontal straight line, the untransformed data may be safely analysed. If, however, the differences tend to increase proportionally with the magnitude of the observations, a log transformation of all the raw data is indicated. If the differences are related to the mean response in some other way, or if the scatter of the points increases (or decreases) with the mean, some other transformation or a non-parametric approach (see next chapter) would probably be advisable.

**Proportions**
Just as it is possible to calculate a standard error as a measure of the

precision of a sample mean in estimating the true population mean, so it is also possible to calculate a standard error for a sample proportion. Supposing, for example, that we are interested in the proportion of children in a population who show evidence of untreated dental caries. In algebraic terms, as noted in an earlier chapter, this proportion of the total population would be denoted by $\pi$. However, as before, we are looking at a sample rather than at the whole population. Suppose, again, that we take random samples of size $n$ from the population, and that, in any given sample, there will be $r$ children with active caries and therefore $n - r$ without. The sample proportion ($p$) with active caries will then be $r/n$. Now the distribution of $p$ follows a *binomial* distribution, but if our samples are reasonably large, this approximates very closely to the *Normal* distribution. It can be shown mathematically that the standard deviation of $r$ (the number of 'successes' out of $n$ binomial trails) is

$$\sqrt{n\pi(1 - \pi)}.$$

Since the sample proportion $p$ equals $r/n$, and since the proportion is really closely akin to a mean, it follows that

$$SD(p) = \frac{1}{n} SD(r) = \sqrt{\frac{n\pi(1 - \pi)}{n^2}} = \sqrt{\frac{\pi(1 - \pi)}{n}}$$

The standard deviation of $p$, like the standard deviation of $\bar{x}$, is usually called the standard error of the proportion and so

$$SE(p) = \sqrt{\pi(1 - \pi)/n}$$

and since for large samples the distribution of $p$ approximates to the Normal distribution, we can say that 95% of repeated sample proportions will lie within the range

$$\pi \pm 1\cdot96\ SE(p) = \pi \pm 1\cdot96\sqrt{\pi(1 - \pi)/n}$$

**Single sample tests of proportions**

The same sort of argument can be applied to an investigation of the significance of the difference between a sample proportion $p$ and the population proportion $\pi$, as is applied to the difference between a sample and a population mean. For example, suppose in a test of two analgesics, A and B, 100 patients were each given the two drugs on different occasions;

65 patients said they preferred A and 35 preferred B. Is this reasonable evidence to support the statement that for these patients, in general, A is the preferred medicament?

If patients in general showed no particular preference for A or B, the proportion of A preferences, $\pi$, would be $0 \cdot 5$. The null hypothesis to be tested here is therefore that the proportion of all patients of this type who prefer A is $0 \cdot 5$.

If $r$ is the observed number of A preferences out of $n = 100$ patients, then $r = 65$ and

$$\text{SD}(r) = \sqrt{n\pi(1 - \pi)} = \sqrt{100 \times 0 \cdot 5 \times 0 \cdot 5} = 5$$

The SND will then be $(r - n\pi)/\sqrt{n\pi(1 - \pi)} = (65 - 50)/5 = 3$

and since 3 is clearly larger than the 1% critical value of $2 \cdot 58$, this is a highly significant result ($P < 0 \cdot 01$).

In practice, with a small sample, the Normal distribution will provide a more precise approximation to the binomial distribution if a *continuity correction* is applied. This involves subtracting $0 \cdot 5$ from the positive value of the difference between the observed frequency and that to be expected if the null hypothesis were true.

Space does not permit a full explanation of the logic behind this continuity correction, but it has to do with the fact that whereas a Normal distribution curve presents as the plot of a smooth continuous variable, the binomial distribution looks more like a pyramidal staircase, consisting of small horizontal steps each taking up one whole unit on the x-axis, followed by a vertical 'jump' to the next step. The continuity correction attempts to find the centre point on the x-axis for each of these 'steps'. Using this correction

$$\text{SND} = (|r - n\pi| - 0 \cdot 5)/\sqrt{n\pi(1 - \pi)}$$

ie $(|65 - 50| - 0 \cdot 5)/\sqrt{100 \times 0 \cdot 5 \times 0 \cdot 5} = 2 \cdot 9$

where $|r - n\pi|$ is called the modulus of $r - n\pi$ and indicates that the contents of the vertical lines should take a positive value; in other words, the smaller of the two values should be subtracted from the larger.

Notice that the SND can be calculated equivalently from the proportion:

$\text{SND} = (p - \pi)/\sqrt{\pi(1 - \pi)/n}$
  or $(p - \pi - 1/2n)/\sqrt{\pi(1 - \pi)/n}$ with the correction
  $= (0 \cdot 65 - 0 \cdot 5)/\sqrt{0 \cdot 5 \times 0 \cdot 5/100}$
      or $(0 \cdot 65 - 0 \cdot 5 - 0 \cdot 005)/\sqrt{0 \cdot 5 \times 0 \cdot 5/100}$
  $= 2 \cdot 9$ as before.

## Comparing two independent sample proportions

Once again, as with the test of a single proportion, advantage is taken of the fact that for large samples the binomial distribution approximates closely enough to the Normal distribution.

Given two sample proportions $p_1$ and $p_2$

$$\text{var}(p_1) = \pi(1 - \pi)/n_1 \text{ and } \text{var}(p_2) = \pi(1 - \pi)/n_2$$
$$\text{and } \text{var}(p_1 - p_2) = \pi(1 - \pi)/n_1 + \pi(1 - \pi)/n_2,$$
$$\text{then } \text{SE}(p_1 - p_2) = \sqrt{\pi(1 - \pi)(1/n_1 + 1/n_2)}$$

Since in this case we do not know $\pi$, we have to replace it by the best single estimate available. This is the proportion $p$, obtained by pooling the two samples

$$p = (r_1 + r_2)/(n_1 + n_2)$$

where $r_1$ and $r_2$ are, as before, the number of subjects in the two samples with the characteristic we are interested in. Then

$$\text{Est.SE}(p_1 - p_2) = \sqrt{p(1 - p)(1/n_1 + 1/n_2)}$$

and

$$\text{SND} = (p_1 - p_2)/\text{Est.SE}(p_1 - p_2)$$

Again, a somewhat more accurate estimate of the SND can be obtained by using a continuity correction

$$\text{SND} = (|p_1 - p_2|) - \tfrac{1}{2}(1/n_1 + 1/n_2))/\text{Est.SE}(p_1 - p_2)$$

In an earlier chapter, reference was made to the dangers of arbitrarily assigning a numerical value to a subjective clinical diagnosis, as is done in many indices of periodontal conditions. The temptation (to which many researchers in this field succumb) is then to analyse such data as if they were indeed quantitative, by making use of $t$-tests on sample means. A more realistic approach, which can also give far more revealing results, is to analyse the proportions of each index grade in the population, using either the techniques described above, or perhaps the chi-squared test, which will be the subject of the next chapter. Tests of significance which may be used when data do not conform to, and cannot be transformed to conform to any of the recognised probability distributions will also be described.

## Practical example

The number of patients visiting the dental departments of two District General Hospitals in a total of 14 equal sample time periods is

Hospital A:    8  12  7  15  9  10  11  13  10  14  9  12  11  12
Hospital B:    11  6  8  12  7  6  9  10  8  10  7  12  11  12

Does one department see significantly more patients than the other?

# 6

# Chi-squared and distribution-free tests

A discussion of basic parametric statistical tests of sample proportions and frequencies is concluded with a description of the chi-squared test. The treatment of distribution-free or non-parametric data is then illustrated with examples of the sign test and Wilcoxon's two-sample rank test.

## Comparing two independent sample proportions

At the end of the last chapter, a method of assessing the difference between two independent sample proportions was described, and an example will now be given. However, the significance of the difference can also be investigated using the chi-squared ($\chi^2$) test, as the subsequent example demonstrates.

## Example

Slack *et al.* (*BDJ* 1971; **130**: 154–158) compared the effects of using two dentifrices, A and B, in children over a period of 3 years. Of the 423 children initially in group A, 163 withdrew over the following 3 years, leaving 260 available at the end for examination, whereas of 408 who started in group B, 119 withdrew, leaving 289.

Is there evidence here to indicate that the proportions withdrawing were dependent on the dentifrice?

The null hypothesis ($H_o$) to be tested is that, in general, the proportion of children in group A who would withdraw ($\pi_a$) is the same as the proportion in group B who would withdraw ($\pi_b$). That is, the observed difference between the sample proportions is due only to sampling error.

ie $$\pi_a = \pi_b = \pi \text{ (for convenience)}$$

If the null hypothesis is true, then $p_a$ (163/423 = 0·385) and $p_b$ (119/

48

408 $= 0 \cdot 292$) are two separate estimates of $\pi$. Combining these two to give a single best estimate of $\pi$ gives

$$p = \frac{(163 + 119)}{(423 + 408)} = 0 \cdot 339$$

$$\text{var}(p_a - p_b) = \text{var}(p_a) + \text{var}(p_b)$$
$$= \frac{\pi(1 - \pi)}{n_a} + \frac{\pi(1 - \pi)}{n_b}$$

$$\text{Est.var}(p_a - p_b) = \frac{0 \cdot 339 \times 0 \cdot 661}{423} + \frac{0 \cdot 339 \times 0 \cdot 661}{408}$$

$$\text{SE}(p_a - p_b) = \sqrt{0 \cdot 339 \times 0 \cdot 661 \left(\frac{1}{423} + \frac{1}{408}\right)}$$
$$= 0 \cdot 0328$$

$$\text{SND} = \frac{p_a - p_b}{\text{SE}(p_a - p_b)} = \frac{0 \cdot 385 - 0 \cdot 292}{0 \cdot 0328}$$
$$= 2 \cdot 84$$

This result is highly significant: $P < 0 \cdot 01$. The null hypothesis ($H_o$) is thus rejected.

$$95\% \text{ CI for } \pi_a - \pi_b = 9 \cdot 3\% \pm 1 \cdot 96 \times 3 \cdot 28\%$$
$$= 2 \cdot 8\% \text{ to } 15 \cdot 8\%$$

It is therefore reasonable to assume that dentifrice A is less popular than dentifrice B, and in trials of this type, the proportion of children in group A who withdraw is likely to be between about 3% and 16% more than the proportion in group B who withdraw.

### The chi-squared test (Greek letter, capital chi: $\chi$)

This test is used to assess quantitatively whether a set of frequencies follows a particular distribution, ie whether *observed* frequencies differ significantly from those *expected* on the basis of some specified theory or hypothesis. It is quick and easy to use, but the results obtained can often fail to provide the full analysis and description that the data may warrant.

### Example

The data from the previous example are used here in order to demonstrate the equivalence of the significance tests.

First a *2 × 2 contingency table* is constructed:

|  | Group | | |
|  | A | B | Total |
|---|---|---|---|
| No. withdrawing | 163 | 119 | 282 |
| No. remaining | 260 | 289 | 549 |
| Total | 423 | 408 | 831 |

The null hypothesis is the same as before, stating, in effect, that the risk that a subject will withdraw from the trial is independent of the dentifrice used.

If $H_0$ is true, the best estimate of the proportion withdrawing is

$$282/831 = 0 \cdot 3394$$

and the expected number of withdrawals from group A would be

$$423 \times 0 \cdot 3394 = 143 \cdot 55$$

Similarly, if $H_0$ is true, the expected number of withdrawals from group B would be

$$408 \times 0 \cdot 3394 = 138 \cdot 48$$

It is therefore now possible to construct a table of *expected* frequencies, similar to the one already prepared of *observed* values:

| | Group | | |
|---|---|---|---|
| | A | B | Total |
| No. withdrawing | 143·5 | 138·5 | 282·0 |
| No. remaining | 279·5 | 269·5 | 549·0 |
| Total | 423·0 | 408·0 | 831·0 |

The general formula for calculating $\chi^2$ is

$$\chi^2 = \Sigma \frac{(O - E)^2}{E}$$

where $O$ = observed value in each cell and $E$ = the expected value.

Thus,
$$\chi^2 = \frac{(163 - 143·5)^2}{143·5} + \frac{(119 - 138·5)^2}{138·5} +$$
$$\frac{(260 - 279·5)^2}{279·5} + \frac{(289 - 269·5)^2}{269·5}$$
$$= 8·17 \text{ on } 1 \text{ df}$$

Notice that the formula $\chi^2 = \Sigma [(O - E)^2/E]$ indicates how closely the observed and expected frequencies agree. If $O - E$ is small, the observed frequencies must be close to the frequencies which would be expected if the null hypothesis were true. Thus, small values of $\chi^2$ imply close agreement between the data and the null hypothesis, while large values of $\chi^2$ imply that the null hypothesis is probably wrong. The values of $\chi^2$ which can be regarded as 'small' or 'large' depend on the number of cells in the table, or, more precisely, the number of independent comparisons of $O$ with $E$. This number is called the *degrees of freedom*, and for a table with $r$ rows and $c$ columns can be shown to be $(r - 1)(c - 1)$. In this example, $r = 2$ and $c = 2$, so there is $(2 - 1)(2 - 1) = 1$ degree of freedom. With one degree of freedom, $SND^2 = \chi^2$. Thus, apart from rounding errors, $8·17$ is the square of an SND value of $2·84$, and the result is therefore, as before, highly significant. The critical values of $\chi^2$ are tabulated in Appendix 1, giving values for various levels of significance (for example $P = 0·05$, $P = 0·01$, $P = 0·001$) and degrees of freedom. If the calculated value of $\chi^2$ is greater than the tabulated critical value, the null hypothesis is rejected at the given level of significance.

The above was an example using a $2 \times 2$ contingency table. Chi-squared tests may be carried out using much larger contingency tables (3 × 4, 4 × 5) and even multidimensional tables, but the larger the tables, the less easy it becomes to interpret the result. Before looking at an example involving a larger table, it should be noted that:

(1) The formula for $\chi^2$, $\Sigma[(O - E)^2/E]$, is only valid for comparing frequencies; it will not, for example, work on percentages.

(2) If the smallest expected frequency in a 2 × 2 table is fairly low, a continuity correction should be applied:

$$\chi^2 = \Sigma[(|O - E| - \tfrac{1}{2})^2/E]$$

(3) If the smallest expected frequency is about 3 or less, $\chi^2$ may give an unreliable result.

### Example using a 3 × 4 contingency table

These data are taken from the report of a dental health campaign (*BDJ* 1967; **123:** 535–536).

|  | Oral hygiene | | | | |
| School | G | F+ | F− | B | Total |
| --- | --- | --- | --- | --- | --- |
| Below average | 62 | 103 | 57 | 11 | 233 |
| Average | 50 | 36 | 26 | 7 | 119 |
| Above average | 80 | 69 | 18 | 2 | 169 |
| Total | 192 | 208 | 101 | 20 | 521 |

These are, of course, the observed frequencies. The expected frequencies are now calculated on the hypothesis that (for example) the same proportion of children with 'good' oral hygiene attended 'below average' schools as did those children with 'fair +', 'fair −' and 'bad' oral hygiene.

The exact proportion of 'good' oral hygiene children is not known, so we have to use the observed proportion, which is 192/521 = 0·3685. Thus, among the 233 children attending 'below average' schools, the *expected* frequency of 'good' oral hygiene is

$$(192/521) \times 233 = 85\cdot9$$

In general, the expected frequency for each cell in the table can be calculated from the formula (row total × column total)/grand total.

## Expected frequencies

| School | Oral hygiene | | | | Total |
|---|---|---|---|---|---|
| | G | F+ | F− | B | |
| Below average | 85·9 | 93·0 | 45·2 | 8·9 | 233·0 |
| Average | 43·9 | 47·5 | 23·1 | 4·6 | 119·1 |
| Above average | 62·3 | 67·5 | 32·8 | 6·5 | 169·1 |
| Total | 192·1 | 208·0 | 101·1 | 20·0 | 521·2 |

$$\chi^2 = \frac{(62 - 85 \cdot 9)^2}{85 \cdot 9} + \frac{(103 - 93 \cdot 0)^2}{93 \cdot 0} +$$
$$\frac{(57 - 45 \cdot 2)^2}{45 \cdot 2} + \ldots + \text{and so on}$$
$$= 6 \cdot 6 + 1 \cdot 1 + 3 \cdot 1 + 0 \cdot 5 + \ldots +, \text{etc}$$
$$= 31 \cdot 4 \text{ on 6df}$$

The tabulated critical value of $\chi^2$ on 6df for $P = 0 \cdot 001$ is $22 \cdot 46$, so the null hypothesis must be rejected. There is clearly a significant association between oral hygiene levels and school status. However, what this test does *not* tell us is where the significance lies. It could be only between two grades of oral hygiene, or between only two types of school, rather than as a gradient involving all three school classifications and all four oral hygiene grades. It would therefore be perfectly possible for 'bad' oral hygiene to be equally prevalent in both 'below average', 'average', and 'above average' schools, but the result of the test could still be significant, due to differences in the prevalence of 'good' oral hygiene. More than one author has also fallen into the trap of assuming that a significant chi-squared result must be significant in the way he or she was expecting. In this example, we might assume that 'significance' implies that the trend is for oral hygiene to be generally 'better' in above average schools; however, a significant result could also be generated by the reverse. The figures themselves have to be very carefully assessed before conclusions are drawn from chi-squared tests concerning the form or direction of the association. It is usually a good idea to calculate the percentage distributions; here we would need to calculate the percentage of children in 'below average' schools who had 'good',

'F +', 'F −', and 'bad' oral hygiene and compare these with the percentages for 'average' and 'above average' schools.

Before leaving chi-squared tests, it should also be noted that with 2 × 2 contingency tables, the test does not provide information about the two values of the proportions, the difference between them, or the standard error of this difference. Consequently, the confidence interval for the true difference between the proportions cannot be calculated. It is therefore less easy to relate statistical to clinical significance when using this test, as compared to the SND test of independent sample proportions.

## Non-parametric tests

Sometimes, the data to be investigated do not follow any recognised distribution pattern, no matter how zealously we try to transform them. Naturally, this severely limits the predictions that can be made about how such data are likely to interrelate, since we have no mathematical guidelines to aid us. Analysis of these data is still possible, but only by using what are called distribution-free or non-parametric methods of analysis. Most non-parametric tests are based on ranking observations in order of magnitude and testing these rankings rather than the actual values of the observations. There are a large number of non-parametric tests, mostly named after their proposers, but relatively few are, or need to be, in common use. All this tends to make the whole subject of distribution-free analysis confusing to the amateur statistician.

These tests are particularly appropriate not only for data which are markedly non-Normal, but also for quantitative data such as periodontal indices, where a subjective severity grading is assigned an arbitrary numerical score, ie the severity is ordered but not quantified. These tests are also quite often used when a quick and easy analysis is required, as many of them are very simple to apply. The *sign test*, for example, is used to analyse paired data.

### Example

In a study of two treatments, A and B, to control plaque formation, subjects are paired according to age, sex, and plaque score before the trial. One subject is randomly allocated to receive treatment A and the other receives treatment B. If, after treatment, the plaque score of A was higher than B in one individual pair, the result might be recorded as ' −', while if B was higher than A it would be recorded as ' +'. If A = B, the result would be a zero difference and eliminated from further analysis. Supposing that as a result out of 30 non-zero pairs, 18 were recorded as ' +' and 12 as ' −'.

If there was really no difference in the effectiveness of the treatments, the proportion of ' + 's should be 0·5, and the test is therefore to see if the actual proportion of ' + 's differs significantly from 0·5. The method used is the test of proportions we have already described in the previous chapter. In this example, $p = 18/30 = 0·6$, and

$$SE(p) = \sqrt{0·5(1 - 0·5)/30} = 0·0913.$$

The SND is then

$$(0·6 - 0·5)/0·0913 = 1·09,$$

which is clearly not significant.

Notice that although the data in this example are non-parametric, the proportion of 'positive' scores, which was the statistic tested, follows a binomial distribution. The test is obviously crude, but it is quick, easy, and reliable.

### Independent samples: Wilcoxon's two-sample rank test
Reference has been made to the confusing number of non-parametric tests available; however, it is fairly safe to say that in circumstances where you would use a $t$-test if the data followed, or could be transformed to follow a Normal distribution with constant variance, you can use the following Wilcoxon test if the data cannot be transformed.

Suppose a sample of $n_1$ observations is to be compared with a sample of $n_2$ observations. Let $n_1$ be the smaller sample:

$$n_1: 3\ 4\ 3\ 3\ 5 \quad \text{(Group A)}$$
$$n_2: 2\ 3\ 3\ 5\ 6\ 4 \quad \text{(Group B)}$$

### Method
(1) All the $(n_1 + n_2)$ observations are ranked in a single array, so that the two samples can still be identified:

| 2 | 3 | 3 | 3 | 3 | 3 | 4 | 4 | 5 | 5 | 6 | – Variables |
|---|---|---|---|---|---|---|---|---|---|---|---|
| 1 | 4 | 4 | 4 | 4 | 4 | 7·5 | 7·5 | 9·5 | 9·5 | 11 | – Ranks |
| B | A | A | A | B | B | A | B | A | B | B | – Groups |

Ranking first would obviously be the '2' value in B. However, ranking second would be equally the three values of '3' in A and the two values of '3' in B. Since between them they will share the second to sixth rankings, they are all assigned the 'median' rank of fourth. The seventh and eighth positions are then occupied by the two '4' values, one from each sample, and these are given the median ranking of $7 \cdot 5$. The two '5' values are similarly ranked $9 \cdot 5$, and the final '6' value is ranked 11th.

(2) The ranks of the observations in the two sample are added up separately.

(3) The smaller sum of these two ranks is designated 'T'.

(4a) If $n_1$ is greater than 2, and $n_1 + n_2$ is greater than 30:

Calculate SND $= (T - \mu)/SE(T)$

Where $\mu = n_1(n_1 + n_2 + 1)/2$

And SE(T) $= \sqrt{n_2\mu/6}$

If the calculated value of the SND exceeds $1 \cdot 96$, $H_o$ is rejected.

(4b) If $n_1 + n_2$ is less than 30, then T is calculated as before, but in this case it has to be compared with critical value tables; these can be found in many comprehensive statistical texts. To make life difficult, these tables work in the opposite way to the SND or $t$-distribution tables we have used up to now, ie $H_o$ is rejected if the calculated value of T is *less* than the critical value!

### Example

The following are waiting times in days for dental treatment in two institutions, A and B.

A: 1, 5, 15, 7, 42, 13, 8, 35, 21, 12, 12, 22, 3, 14, 4, 2, 7, 2.

$$n_2 = 18$$

B: 4, 9, 6, 2, 10, 11, 16, 18, 6, 0, 9, 11, 7, 11, 10.

$$n_1 = 15$$

| Time | Institution | Rank | Time | Institution | Rank |
|------|-------------|------|------|-------------|------|
| 0 | B | 1 | 10 | B | 18·5 |
| 1 | A | 2 | 10 | B | 18·5 |
| 2 | A | 4 | 11 | B | 21 |
| 2 | B | 4 | 11 | B | 21 |
| 2 | A | 4 | 11 | B | 21 |
| 3 | A | 6 | 12 | A | 23·5 |
| 4 | A | 7·5 | 12 | A | 23·5 |
| 4 | B | 7·5 | 13 | A | 25 |
| 5 | A | 9 | 14 | A | 26 |
| 6 | B | 10·5 | 15 | A | 27 |
| 6 | B | 10·5 | 16 | B | 28 |
| 7 | A | 13 | 18 | B | 29 |
| 7 | A | 13 | 21 | A | 30 |
| 7 | B | 13 | 22 | A | 31 |
| 8 | A | 15 | 35 | A | 32 |
| 9 | B | 16·5 | 42 | A | 33 |
| 9 | B | 16·5 | | | |

Sum of A ranks $= 2 + 4 + 4 + 6 + 7·5 + \ldots + 32 + 33 = 324·5$

Sum of B ranks $= 1 + 4 + 7·5 + 10·5 + \ldots + 28 + 29 = 236·5$

$$T = 236·5$$

$$\mu = 15(15 + 18 + 1)/2 = 255$$

$$SE(T) = \sqrt{(18 \times 255)/6} = \sqrt{765} = 27·66$$

$$SND = (236·5 - 255)/27·66 = -0·67$$

This is clearly not statistically significant at the 5% level. There is no strong evidence that waiting times are any longer at one institution than at the other.

In the next chapter, consideration will be given to problems where two variables interact, and we are interested in the degree of association between the two. For example, how does the prevalence of fluorosis in populations relate to the fluoride level in the water supply?

## Practical example

In an investigation of 676 adult orthodontic patients, of whom 192 are men and 484 are women, it is found that 47 of the men and 80 of the women have undergone orthognathic surgery (data from Khan R. S., personal communication). Is there a significant difference in the proportions of men and women undergoing this type of surgery? Answer this question using:

(1) The chi-squared test
(2) An SND test on the difference between the two proportions.

# 7

# Analysing the association between two variables

Techniques of linear regression and correlation analysis are discussed and illustrated by means of dentally-related examples.

## Linear regression

Up to now we have dealt with one variable at a time (for example DMF), and compared changes in that variable in different circumstances (eg in different parts of the country). However, the interest frequently lies not just in the way a variable may change in isolation, but rather in the way in which it may apparently be influenced by, or be associated with, another variable. For example, up to a certain age we get taller as we get older, so what is the mean increase in height in a given population for each year of life? To determine this we obviously have to record both the age and the height of each and every member of a random sample drawn from the population under review. The next step is to draw a *scatter diagram* of the results. This is done by plotting each pair of observations on a graph where, conventionally, the horizontal or 'x' axis represents the explanatory or *independent variable* (ie age), and the vertical or 'y' axis represents the response or *dependent variable* (ie height). In other words, for any given age we are measuring differences in height rather than the other way round.

This scatter diagram preparation is an important part of the investigation, since, among other things, just 'eyeballing' the plotted data can often indicate that there is no strong association between the two variables and that further analysis is therefore probably unnecessary. The points thus plotted in the scatter diagram may or may not tend to follow a linear pattern, depending on the degree and nature of the association between the two variables. For example, suppose we were investigating a new filling material whose resistance to stress, as measured by its ability to withstand crushing pressure, was apparently related in some way to lapsed time after placement, measured as lapsed time after the maximum manipulative time recommended by the manufacturers. We first want to know the relationship between lapsed time and crushing strength, and

**Table I**   Paired lapsed time from placement (x) and crushing strength (y) for specimens of hypothetical filling material

| x | y |
|---|---|
| 1·0 | 0·250 |
| 2·0 | 0·875 |
| 3·0 | 0·750 |
| 4·0 | 1·625 |
| 5·0 | 2·000 |
| 6·0 | 2·125 |
| 7·0 | 2·250 |
| 8·0 | 3·250 |

secondly we need to be able to predict the likely crushing strength (measured in any convenient units) at any given time after placement. Table I therefore gives the crushing strength (the dependent variable, $y$) at each of eight one-minute intervals after placement (the independent variable, $x$). Of course in reality we would not be content with just eight pairs of readings, but for the sake of simplicity we ask for your indulgence in this respect. The next step is to prepare the scatter diagram of these pairs of data. The result is shown in figure 1. From this it can be seen that the points do indeed approximate to a straight line, and it is not unreasonable to suggest that such a linear relationship may exist between crushing

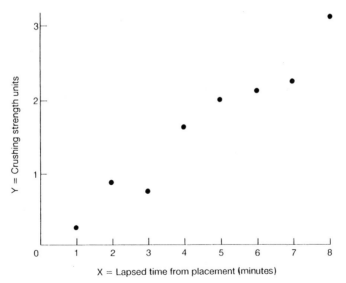

**Fig. 1** Scatter diagram of data from Table I.

strength and lapsed time, at least over the period for which we have data. The general formula for such a straight line is

$$y = a + bx$$

where $b$ is the slope or gradient of the line (that is, the tangent of the angle between the line and the horizontal). The constant $a$ is the value of $y$ when $x = 0$, ie the point at which the line crosses the $y$ axis.

For the purposes of this analysis, however, we use the formula

$$y = \alpha + \beta x$$

where $\beta$ still corresponds to $b$, and $\alpha$ to $a$, but the Greek letters refer to the 'population' of points, while $a$ and $b$ are derived from a sample which is being used to provide an estimate of the population parameters $\alpha$ and $\beta$.

Thus, our problem is to find sample estimates $a$ and $b$ of the true population's $\alpha$ and $\beta$, so that the resulting calculated line provides the best estimate of the true population line. The method used to determine the best fitting line involves calculating the vertical distance ($d$) of each point from the line, in such a way that $\Sigma d^2$ is minimised. This is known as fitting a line by the *method of least squares* (fig. 2). It can be shown mathematically

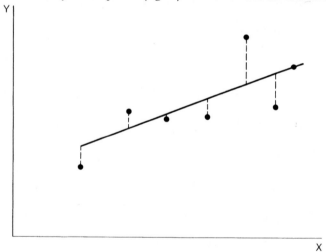

**Fig. 2** Fitting a straight line by the method of least squares.

that in order to fulfil this requirement

$$b = \frac{\Sigma(x - \bar{x})(y - \bar{y})}{\Sigma(x - \bar{x})^2}$$

and that

$$a = \bar{y} - b\bar{x}.$$

At this point we will introduce a shorthand algebraic notation which avoids some rather unwieldy formulae. At the end of the second chapter we noted that

$$\Sigma(x - \bar{x})^2 = \Sigma x^2 - (\Sigma x)^2/n.$$

From now on we will refer to this particular formula as X. Therefore, whenever you see X you will know that it means $\Sigma(x - \bar{x})^2$ or, equivalently, $\Sigma x^2 - (\Sigma x)^2 n$. In the same way, logically

$$\Sigma(y - \bar{y})^2 = \Sigma y^2 - (\Sigma y)^2/n$$

and this we will call Y.

There is now one new formula to consider:

$$\Sigma(x - \bar{x})(y - \bar{y}) = \Sigma xy - (\Sigma x)(\Sigma y)/n$$

and this we will denote as Z. Using this new shorthand, we can now say that the formula for $b$ is Z/X. In order to avoid possible confusion, it should be noted that in many statistical textbooks X, Y and Z are referred to as $S_{xx}$, $S_{yy}$ and $S_{xy}$, respectively.

The constant $a = \bar{y} - b\bar{x}$ is known as the *intercept*, and the constant $b = $ Z/X is known as the *estimated regression coefficient*; $b$ is also known (as mentioned above) as the slope of the line, since it measures the average increase in $y$ per unit increase in $x$.

We now have all the information we need to calculate the regression line of crushing strength on lapsed time, and you should note that with real data these calculations would be made by means of a computer rather than, as here, by hand:

| $x$ | $x^2$ | $y$ | $y^2$ | $xy$ |
|---|---|---|---|---|
| 1·0 | 1·0 | 0·250 | 0·0625 | 0·25 |
| 2·0 | 4·0 | 0·875 | 0·7656 | 1·75 |
| 3·0 | 9·0 | 0·750 | 0·5625 | 2·25 |
| 4·0 | 16·0 | 1·625 | 2·6406 | 6·50 |
| 5·0 | 25·0 | 2·000 | 4·0000 | 10·00 |
| 6·0 | 36·0 | 2·125 | 4·5156 | 12·75 |
| 7·0 | 49·0 | 2·250 | 5·0625 | 15·75 |
| 8·0 | 64·0 | 3·250 | 10·5625 | 26·00 |
| 36·0 | 204·0 | 13·125 | 28·1718 | 75·25 |
| $= \Sigma x$ | $= \Sigma x^2$ | $= \Sigma y$ | $= \Sigma y^2$ | $= \Sigma xy$ |

$$n = 8 \quad \bar{y} = 13\cdot125/8 = 1\cdot6406 \quad \bar{x} = 36/8 = 4\cdot50$$
$$X = 204 - (36^2/8) = 42$$
$$Y = 28\cdot1718 - (13\cdot125^2/8) = 6\cdot6386$$
$$Z = 75\cdot25 - (36\times13\cdot125/8) = 16\cdot1875$$
$$\text{So } b = Z/X = 16\cdot1875/42 = 0\cdot3854$$

and $a = \bar{y} - b\bar{x} = 1\cdot6406 - (0\cdot3854 \times 4\cdot5) = -0\cdot09375$

Best fitting straight line:

$$y = a + bx$$
$$= -0\cdot09375 + 0\cdot3854x$$

So if $x = 0$,     $y = -0\cdot094$
$x = 8$,     $y = 2\cdot99$
$x = 4\cdot5$ $(\bar{x})$, $y = 1\cdot64$ $(\bar{y})$

This last pair provides a useful check on the calculation of the regression line, since the line should pass through the point $(\bar{x}, \bar{y})$. It is possible, if we so wish, to draw this calculated line in on the scatter diagram. An important warning must be given here. It is very unwise to use this formula to predict values of $y$ outside the observed range of $x$. Just because, in our example, crushing strength has followed a close straight-line relationship with time up to nearly 8 minutes from placement, we cannot therefore assume that this relationship will continue. Indeed, if it did we would soon end up with a filling material harder than diamonds! Obviously, the relationship is soon going to tail off, with increase in crushing strength per unit of time becoming less and less, until it becomes constant, with no further increase. The use of this formula to calculate $y$ for, say, $x = 10$ is therefore speculative. It should also be noted that if the relationship between the variables follows a curve rather than a straight line, it is often still possible to compute a best-fitting line, provided the data can be transformed (frequently by the use of log transformation) to a straight-line association. Thus, instead of plotting $y$ against $x$, we might plot log $y$ against $x$, or $y$ against log $x$, or log $y$ against log $x$. This, incidentally, is the fourth example of the use of a log transformation, referred to in chapter 5.

It can be shown that the variance of $b$ is given by the formula:

$$\text{Var}(b) = \frac{Y - (Z^2/X)}{(n - 2)X}$$

So 95% confidence limits for $\beta$ will be

$$b \pm t(0 \cdot 05, n - 2) \sqrt{\frac{Y - (Z^2/X)}{(n - 2)X}}$$

Since $t(0 \cdot 05, 6) = 2 \cdot 45$, then 95% CI for $\beta$ are

$$0 \cdot 3854 \pm 2 \cdot 45 \sqrt{\frac{6 \cdot 6386 - (16 \cdot 1875^2/42)}{6 \times 42}}$$

$$= 0 \cdot 3854 \pm 2 \cdot 45 \times 0 \cdot 03982$$
$$= 0 \cdot 3854 \pm 0 \cdot 09758$$
$$= 0 \cdot 2878 \text{ to } 0 \cdot 4830$$

We would therefore expect the true slope of the population regression line to be not less than $0 \cdot 2878$, but not more than $0 \cdot 4830$, or, alternatively, that the increase in crushing strength per minute of lapsed time from mixing would be unlikely to be less than $0 \cdot 2878$ or more than $0 \cdot 4830$. This is quite a wide range, but remember that we have based our estimate on only eight pairs of readings. The larger the sample size, the narrower these limits would become.

On average, the crushing strength increases by $0 \cdot 4$ units (95% CI, $0 \cdot 3$ to $0 \cdot 5$) per minute.

## Correlation

So far we have been dealing with a situation where our two variables show some reasonable linear association with each other. However, such relationships are usually not so clear-cut. For example, consider figure 3. Scatter diagram A shows a definite linear relationship between $x$ and $y$, but diagram B, although showing a tendency for $y$ to increase as $x$ increases, suggests that other factors may also be involved. How, then, can the actual degree of association be measured?

This is done by calculating a *correlation coefficient*, which is usually designated $r$, where

$$r = Z/\sqrt{XY}$$

The properties of $r$ are:

(1) The sign ($+$ or $-$) of $r$ is always the same as that of Z, and because the regression coefficient = Z/X, $r$ has the same sign as $b$.

(2) $r$ can never be less than $-1$ or greater than $+1$ (if it is, your arithmetic is at fault!)

(3) If $r = \pm 1$, the points lie exactly on a straight line (ie $r = +1$ indicates perfect positive correlation, where $y$ increases as $x$ increases, and $r = -1$ indicates perfect negative correlation, where $y$ decreases as $x$ increases).

(4) If $r = 0$, this *may* mean that there is no association between $x$ and $y$, but it can also mean that an association exists, but that it is non-linear. Figure 4 shows two scatter diagrams where $r = 0$.

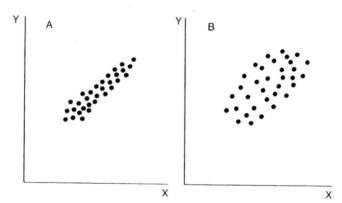

**Fig. 3** Distributions showing (A) relatively high and (B) relatively low correlation.

## Example
Using the lapsed time/crushing strength data

$$r = Z/\sqrt{XY} = \overline{16 \cdot 1875/\sqrt{42 \times 6 \cdot 6386}}$$
$$= 0 \cdot 9694$$

Here, as we would expect from the scatter diagram, we have a strongly positive association. The correlation coefficient is a measure of the closeness of the association between two variables.

## The variability of $y$
The linear regression technique used so far has only provided a prediction of $y$ as it relates to $x$. However, if the points in the scatter diagram are not closely associated with the line (as in B in fig. 3), even if the trend is linear, then other factors may be acting on $y$. We may therefore wish to know the extent to which $x$ can explain observed variations in $y$.

As we have seen, the total variability of $y$ is measured by $\Sigma(y - \bar{y})^2$ (ie Y). What fraction of this *total sum of squares* can *not* be explained by the regression of $y$ on $x$? In other words, what is the variability of the points about the regression line?

It can be shown mathematically that this variability about the line, which is known as the *sum of squared residuals*, is

$$Y - Z^2/X$$

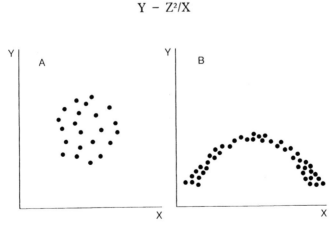

**Fig. 4** Distributions with correlation coefficient ($r$) equal to zero.

Thus, the variability of $y$ which is 'explained' by $x$ is

$$\text{Total SSq.} - \text{SSq. Residual}$$

$$= Y - Y - (Z^2/X) = Z^2/X$$

and so the fraction of the total variability of $y$ explained by $x$ is

$$Z^2/XY \text{ which is (rather surprisingly!) } r^2$$

Thus, in the crushing strength/lapsed time example, 94% of the total variability of crushing strength can be explained by changes in the lapsed time from mixing. Suppose, however, that in comparing DMF with sugar consumption, the correlation coefficient was found to be only $0 \cdot 7$. Then $r^2 = 0 \cdot 49$ and so only half the total variability of DMF scores can be explained by changes in sugar consumption. Other factors would be involved in the remaining 51%.

If the sum of squared residuals described above is divided by the associated number of degrees of freedom (in this case $n - 2 = 6$), the resulting *residual mean square* is seen to be part of the formula for the variance of $b$, which we used earlier in calculating $t$ and the 95% confidence interval for $\beta$.

$$t = (b - 0)/(s\sqrt{X}) \text{ where:}$$

$$s = \sqrt{\text{residual mean square}} = \sqrt{(Y - (Z^2/X))/(n - 2)}$$

### Spearman's rank correlation coefficient

The coefficient of linear correlation described above assumes that in the population from which the pairs of observations were taken, the distribution of $y$ for a given value of $x$ is Normal, with a constant variance (that is, the variance of $y$ for a given value of $x$ does not change if $x$ changes). If these conditions cannot be met, even by data transformation, then Spearman's rank correlation, a non-parametric equivalent, may be invoked. As its name implies, and like many other non-parametric tests, it is based on the rank of the variables rather than their actual values.

### Example

Suppose that in 15 patients under observation in a periodontal clinic, a note is made of the change in the number of teeth per patient showing pocketing in excess of 3 mm after a period of 2 months, with the object of seeing whether the number of pockets tends to increase or decrease significantly during the 2-month period. $(x)$ is then the number of teeth per patient with pockets in excess of 3 mm at the initial examination, and $(y)$ is the number of teeth per patient with pockets 2 months later. These data are shown in Table II.

The observations in each column are then ranked in order of magnitude, and the second ranking subtracted from the first to give the difference $(d)$, as shown in Table III. If there are no 'tied ranks' in either column, that is more than one observation with the same ranking, then Spearmans rank correlation coefficient $(r_s)$ is defined as

$$1 - [(6\Sigma d^2)/(n^3 - n)]$$

However if, as here, there are tied ranks present, the following procedure is followed.

**Table II**  Number of periodontal pockets deeper than 3 mm per patient at commencement $(x)$ and two months after commencement $(y)$ of investigation

| Patient | x | y | Patient | x | y |
|---------|---|---|---------|---|---|
| 1 | 3 | 0 | 9  | 2 | 2 |
| 2 | 0 | 2 | 10 | 0 | 0 |
| 3 | 2 | 2 | 11 | 6 | 5 |
| 4 | 0 | 0 | 12 | 0 | 0 |
| 5 | 1 | 1 | 13 | 0 | 4 |
| 6 | 1 | 4 | 14 | 2 | 0 |
| 7 | 4 | 2 | 15 | 2 | 3 |
| 8 | 1 | 1 |    |   |   |

**Table III**  Ranking and difference between ranks $(d)$ of periodontal pocket data given in Table II

| Patient | Rank(x) | Rank(y) | x − y = d | Patient | Rank(x) | Rank(y) | x − y = d |
|---------|---------|---------|-----------|---------|---------|---------|-----------|
| 1 | 13   | 3    | 10·0  | 9  | 10·5 | 9·5  | 1·0   |
| 2 | 3    | 9·5  | −6·5  | 10 | 3    | 3    | 0·0   |
| 3 | 10·3 | 9·5  | 1·0   | 11 | 15   | 15   | 0·0   |
| 4 | 3    | 3    | 0·0   | 12 | 3    | 3    | 0·0   |
| 5 | 7    | 6·5  | 0·5   | 13 | 3    | 13·5 | −10·5 |
| 6 | 7    | 13·5 | −6·5  | 14 | 10·5 | 3    | 7·5   |
| 7 | 14   | 9·5  | 4·5   | 15 | 10·5 | 12   | −1·5  |
| 8 | 7    | 6·5  | 0·5   |    |      |      |       |

First, the number of tied ranks in each column must be noted.
For column $(x)$:

| | | | |
|---|---|---|---|
| 3 | 7 | $10 \cdot 5$ | — Tied ranks |
| 5 | 3 | 4 | — $t$ (number of ties) |
| 10 | 2 | 5 | — $T$, where $T = (t^3 - t)/12$ |

$$T_x = \Sigma T \text{ for } x \text{ column} = 17$$

For column $(y)$:

| | | | | |
|---|---|---|---|---|
| 3 | $6 \cdot 5$ | $9 \cdot 5$ | $13 \cdot 5$ | — Tied ranks |
| 5 | 2 | 4 | 2 | — $t$ |
| 10 | $0 \cdot 5$ | 5 | $0 \cdot 5$ | — $T$ |

$$T_y = \Sigma T \text{ for } y \text{ column} = 16$$

$$\text{Let } A = (n^3 - n)/12 = 280$$

$$\begin{aligned}
\text{Then } r_s &= (2A - T_x - T_y - \Sigma\, d^2)/[2\sqrt{(A - T_x)(A - T_y)}] \\
&= (560 - 17 - 16 - 376)/[2\sqrt{(280 - 17)(280 - 16)}] \\
&= 151/526 \cdot 999 = 0 \cdot 29
\end{aligned}$$

This result shows that there is no strong association between the number of deep pockets per patient at the start and end of the 2-month period, although it has to be remembered that a contributory factor to such a result could well be examiner inconsistency, especially if the examiners involved had not been properly trained and calibrated.

Before leaving this topic, there is a very important point to make concerning the interpretation of correlation and regression data. Just because two variables can be shown to have a high, or even perfect, correlation, this does not mean that the change in one variable is *causing* the change in the other. Association does not of itself imply causation. Thus, to invoke the fact that there is a strong association between national sugar consumption figures and dental caries levels to support the hypothesis that increased sugar consumption causes an increase in dental caries is illogical, since the data equally suggest that increasing levels of dental caries cause an increase in sugar consumption. The correlation coefficient may *generate* hypotheses, but should be merely the first step in the analytical process which seeks to eliminate all the other myriad confounding variables which could be influencing both the levels of sugar consumption and the levels of dental caries.

To do otherwise is to invite ridicule. It could, for example, be easily shown that the whisky consumption in a given community shows a very strong association with the number of dental surgeons in the community. Are we therefore to assume that the dentists are being attracted to the community by the availability of whisky, or that the whisky is being attracted to the community by the dentists? Both ideas are clearly bizarre, and the truth lies in the fact that the overall population of the community is itself strongly associated with both whisky consumption and the number of dental surgeons.

## Practical example

A university lecturer wishes to know if the marks obtained by dental students in a 'rehearsal' for a written professional examination are likely to provide a reasonable prediction of their performance in the 'real thing'. From the previous year, he has therefore recorded for each student the mark obtained in the 'rehearsal' ($x$) and the mark obtained in the professional examination ($y$). These are:

| Student | $x$ | $y$ |
|---------|-----|-----|
| 1 | 15 | 37 |
| 2 | 18 | 46 |
| 3 | 19 | 47 |
| 4 | 24 | 48 |
| 5 | 19 | 52 |
| 6 | 22 | 54 |
| 7 | 26 | 58 |
| 8 | 28 | 61 |
| 9 | 28 | 65 |

(1) What would you expect a student's score to be in the real examination if he obtained a score of 20 in the rehearsal?

(2) What is the correlation coefficient for the two sets of results?

# 8

# Comparison of several groups

Techniques for the comparison of several samples are described. If the response is quantitative, comparison of the sample means is achieved by one-way analysis of variance. If the response is qualitative, comparison of several proportions is achieved by $\chi^2$ tests.

**Comparing several sample means**
In describing the $t$-test, the assumption was made that we were interested in comparing just two independent sample means. However, it is quite often necessary to compare three or more such means. What should the procedure be in such circumstances?

One fairly obvious solution would be to carry out $t$-tests on each and every possible pairing of samples within the group. Thus, if we had three samples A, B, and C, the total pairings AB, BC, and AC could be tested. There are two major objections to this approach. First, the number of possible pairings increases alarmingly as the number of means to be tested increases, so that five means, for example, will yield ten possible pairings. Secondly, it has to be remembered that if we decide to adopt the 5% level of significance, there is a 5% chance in any one of these tests, of obtaining a 'significant' result, if, in fact, no real difference exists. This, incidentally, is known as a Type I error. Clearly, the more $t$-tests which are carried out on the data set, the greater the likelihood of obtaining Type I errors. In the five-sample situation referred to above, with ten possible pairings, the probability that at least one test out of the ten will yield a Type I error could be as high as $0 \cdot 4$ or 40% (ie $1 - 0 \cdot 95^{10}$). In practice, the probability will be somewhat less than 40%, as we are dealing with significance levels where $P$ will be less than $0 \cdot 05$ rather than $P$ equalling $0 \cdot 05$, but the figure will still be high. Also, the ten tests will not be ten *independent* tests.

Some rethinking of the problem is therefore indicated. You will recall that an assumption underlying the $t$-test comparison of two independent sample means is that the variance obtained from each sample ($s_1{}^2$, $s_2{}^2$) is an

70

estimate of the true population variance $\sigma^2$. The best estimate of $\sigma^2$ is therefore obtained by combining $s_1^2$ and $s_2^2$.

We can apply similar logic to the multi-sample situation. Suppose we have $k$ samples and, for simplicity, suppose each has $n$ observations. It follows that all the sample variances ($s_1^2$, $s_2^2$ . . . $s_k^2$) are estimates of the population variance $\sigma^2$, and a combined estimate may be obtained by calculating the weighted average of all the sample variances. This will then be the best estimate of the observations *within* each sample. By comparing more than two samples, however, we are introducing another dimension, because it is not only possible to look at the variances *within* the samples, but also to look at variances *between* the sample means. If the null hypothesis ($H_0$) is true, the variance between the sample means can be predicted, because it should be theoretically $\sigma^2/n$. By calculating $n$ times the *observed* variance of the sample means, we can obtain another estimate of $\sigma^2$. If $H_0$ is true, the estimate of $\sigma^2$ obtained within samples and the estimate obtained between samples should be roughly the same. However, if $H_0$ is false, and there is a genuine difference between one or more of the means, the *within* sample variance will be unaffected (remember that one of the most important criteria for the $t$-test was that the variances of the samples under test had to be the same), but the *between* sample estimate of variance will be large, since it will reflect the real difference between the sample means. A significance test may therefore be devised, which compares the 'within-group' variance and the 'between-group' variance. This is most easily done by calculating the ratio of the two variances. If this ratio is roughly unity, then there is little evidence against $H_0$; however, if it is greater than unity, then there is an increasing chance that $H_0$ is false.

This procedure is known as an 'analysis of variance', and we have already come across the technique when looking at the variability of $y$ in the previous chapter on linear regression and correlation. It is curious, at first sight, that a significance test on sample means is achieved by comparing variances!

Probably the easiest way of explaining the technique, without going into excessive algebraic detail, is to give a simple worked example. Of course in practice all these calculations can be made by computer, with considerable saving in time and mental fatigue!

### Example
Suppose that we have three sets of readings of the number of patients seen in a given period by practitioners in three group practices, and we need to know if the mean number differs significantly between the practices.

The data might be (and notice that the number of observations in each sample does not have to be the same):

| Practice: | A | B | C |
|---|---|---|---|
| | 268 | 387 | 161 |
| | 349 | 264 | 346 |
| | 328 | 423 | 324 |
| | 209 | 254 | 293 |
| | 292 | | 239 |
| | | | |
| $\bar{x}$ | 288·2 | 332·0 | 272·6 |
| $n$ | 5 | 4 | 5 |
| $\Sigma x$ | 1441 | 1328 | 1363 |
| $\Sigma x^2$ | 426899 | 462910 | 393583 |
| $(\Sigma x)^2$ | 2076481 | 1763584 | 1857769 |
| $\mathbf{X}$ | 11602·8 | 22014·0 | 22029·2 |
| $s^2$ | 2900·7 | 7338 | 5507·3 |

The formulae in the calculations which follow may appear clumsy, but the theory behind them is the same as we have already discussed in earlier chapters.

First we calculate the *total sum of squares*. This is

$$(\Sigma x_A^2 + \Sigma x_B^2 + \Sigma x_C^2) - \frac{(\Sigma x_A + \Sigma x_B + \Sigma x_C)^2}{n_A + n_B + n_C}$$

$$= (426899 + 462910 + 393583) - \frac{(1441 + 1328 + 1363)^2}{5 + 4 + 5}$$

$$= 1283392 - (4132^2/14)$$

$$= 63861 \cdot 7143$$

Next, the weighted average of the within-group variances is calculated. This is

$$\frac{(n_A-1)s_A{}^2+(n_B-1)s_B{}^2+(n_C-1)s_C{}^2)}{(n_A-1)+(n_B-1)+(n_C-1)}$$

$$= ((4\times2900\cdot7)+(3\times7338)+(4\times5507\cdot3))/(4+3+4)$$

$$= 55646/11 = 5058\cdot7273$$

Next, the *between-group sum of squares* is calculated:

$$(\Sigma x_A)^2/n_A+(\Sigma x_B)^2/n_B+(\Sigma x_C)^2/n_C) - \frac{(\Sigma x_A+\Sigma x_B+\Sigma x_C)^2}{(n_A+n_B+n_C)}$$

$$= (1441^2/5+1328^2/4+1363^2/5) - \frac{(1441+1328+1363)^2}{5+4+5}$$

$$= 1227746 - (4132^2/14) = 8215\cdot71429$$

Since the total SSq = between-group SSq + within-group SSq, it follows that the within-group SSq = total SSq − between-group SSq: within-group SSq = $63861\cdot7143 - 8215\cdot71429 = 55646\cdot0$

This value is the same as the numerator we obtained when calculating the weighted average of the within-group variances, and provides a useful check on our arithmetic.

As already stated, these calculations would normally be carried out using a computer statistical program, and part of the output from such a program could be presented in an *analysis of variance table:*

| Source of variation | SSq | DF | MSq | VR |
|---|---|---|---|---|
| Between groups | 8215·71 | 2 | 4107·86 | 0·81 |
| Within groups | 55646·00 | 11 | 5058·73 | |
| Total | 63861·71 | 13 | | |

The between- and within-group mean squares (MSq) are obtained by dividing the relevant sums of squares (SSq) by the associated degrees of freedom (DF).

The *variance ratio* (VR = $0\cdot81$) is obtained by dividing the between-groups mean square by the within-groups mean square. This is usually known as the F-ratio, so called after the eminent statistician, R. A. Fisher, who pioneered this work.

The calculated F-ratio value must now be compared with critical values given in a table of the F distribution (*see* Appendix 2), just as we did for $t$ or $\chi^2$ values. However, in this case we are dealing with degrees of freedom for each of the two variance estimates, that of the numerator (in this case 2 df), and that of the denominator (14 df). To look up the critical value we therefore have to select the column of the table headed '2 df' and then trace down it, until we come to the line corresponding to '14 df'. The critical F value lies at the point where the two intersect; in this case, for the 5% level of significance, the critical value is $3 \cdot 98$. Our calculated variance ratio, being less than one, clearly does not exceed this critical value and we would therefore conclude that these data show no strong evidence of a difference between the group practices in the mean case-load per practitioner. It is, of course, informative to also quote the three means and their standard errors, especially as a 'significant' F-ratio does not necessarily imply that all the group means differ significantly from each other:

| | Mean | Standard error |
|---|---|---|
| Group A | $288 \cdot 2$ | $\sqrt{5058 \cdot 73/5} = 31 \cdot 8$ |
| Group B | $332 \cdot 0$ | $\sqrt{5058 \cdot 73/4} = 35.6$ |
| Group C | $272 \cdot 6$ | $\sqrt{5058 \cdot 73/5} = 31 \cdot 8$ |

**Comparison of several sample proportions**

In a study of the effect of fluoride in drinking water on the prevalence of dental caries,[1] a sample of schoolchildren aged 12–14 years was inspected in each of five regions; the results are shown in Table I. This is an example of a 2 × 5 contingency table, of the sort discussed in chapter 6. A simple but crude and inefficient analysis would be to calculate $\chi^2$ with 4 df from the general formula $\Sigma(O - E)^2/E$. This gives $\chi^2 = 80 \cdot 2$ with 4 df, which is highly significant. We would therefore conclude that there is very strong evidence of a difference in prevalence between the five areas.

This result, however, does not do the data justice; there is much more information to be gleaned from it. First, it would be informative to calculate the prevalence of caries (or, equivalently, the proportion of children with zero DMF) in the five areas. Secondly, it would be interesting to see if the proportion carious (or caries free) changes systematically with the fluoride content of the drinking water. The last line in Table I suggests that there

**Table I** Comparison of caries prevalence in areas with varying water fluoride concentrations

| Area | Surrey and Essex | Slough | Harwich | Burnham | West Mersea | Total |
|---|---|---|---|---|---|---|
| F ppm | 0·15 | 0·9 | 2·0 | 3·5 | 5·8 | |
| With caries | 243 | 83 | 60 | 31 | 39 | 456 |
| Caries free | 16 | 36 | 32 | 31 | 12 | 127 |
| Sample-size | 259 | 119 | 92 | 62 | 51 | 583 |
| Percentage caries free | 6·2 | 30·3 | 34·8 | 50·0 | 23·5 | |

**Table II** Algebraic representation of data in Table I

| Group (area) | 1 | 2 | 3. . . . | k | Total |
|---|---|---|---|---|---|
| Explanatory variable (F ppm) | $x_1$ | $x_2$ | $x_3 \cdots$ | $x_k$ | |
| Successes (caries free) | $r_1$ | $r_2$ | $r_3 \cdots$ | $r_k$ | $R(127)$ |
| Failures (caries) | $n_1 - r_1$ | $n_2 - r_2$ | $n_3 - r_3 \cdot$ | $n_k - r_k$ | $N - R(456)$ |
| Sample size | $n_1$ | $n_2$ | $n_3$ | $n_k$ | $N(583)$ |
| Proportion (caries free) successful | $P_1 = r_1/n_1$ | $P_2 = r_2/n_2$ | $P_3 = r_3/n_3$ | $P_k = r_k/n_k$ | $P = R/N$ (0·2178) |

certainly appears to be a trend in the percentage caries free with increasing water fluoride content, but is a linear association statistically significant? Furthermore, if there is a linear trend, do any of the areas differ from the general trend by more than would be reasonably expected by chance?

The answers to these questions can be obtained by decomposing the value of $\chi^2_4$ (ie $\chi^2$ with 4 df) calculated above into two components. The first component, with one degree of freedom, measures the significance (or otherwise) of the linear trend in the proportion caries free with water fluoride content, while the difference between the crude $\chi^2_4$ and the linear trend $\chi^2_1$ will measure the significance of the deviations from the linear trend.

Table II gives the necessary basic algebra for these calculations, together with the equivalent values derived from our example. The crude $\chi^2_{k-1}$ (in our example, $\chi^2_4$) can be calculated either using the formula $\Sigma(O - E)^2/E$ or, alternatively,

$$\frac{\Sigma r^2/n - R^2/N}{P(1 - P)}$$

which in our example is $(41 \cdot 33 - 27 \cdot 67)/0 \cdot 1704 = 80 \cdot 2$ (as before). The component of $\chi^2$ due to linear trend is

$$\chi^2_1 = \frac{N(N\Sigma rx - R\Sigma nx)^2}{R(N-R)[N\Sigma nx^2 - (\Sigma nx)^2]}$$

$$= \frac{583[(583 \times 276 \cdot 9) - (127 \times 803 \cdot 9)]^2}{127 \times 456[(583 \times 2945 \cdot 36) - 803 \cdot 9^2]}$$

$$= 33 \cdot 1$$

which is very much greater than the critical value for the $0 \cdot 1\%$ level of significance. Finally, the deviation from linear trend is measured by $\chi^2_3 = \chi^2_4 - \chi^2_1$, which is $80 \cdot 2 - 33 \cdot 1 = 47 \cdot 1$. This is also highly significant, largely because of the low proportion caries free in West Mersea.

Thus, just as it is possible to compare several sample means using one-way analysis of variance, so several proportions can be compared using the crude $\chi^2$ test. Furthermore, just as a trend in the means can be investigated using linear regression, so the crude $\chi^2$ can be partitioned to test the significance of the linear trend in sample proportions.

The next chapter will be devoted to a consideration of the problems of examiner variability or consistency in clinical dental examinations, and how this may affect the outcome of epidemiological investigations. There is little point in carrying out statistical analysis on data which are flawed as a result of diagnostic inconsistency or examiner bias, yet the recognition given to this problem in published articles tends to be superficial.

## Reference

1 Forrest J. The fluoridation of public water supplies; (a) The dental aspect. *J R Soc Health* 1957; **77**: 344–350.

## Practical example

The following fictitious data provide measures of mean upper central incisor width (mm) in individuals from four different ethnic groups.

| | | Group | |
|---|---|---|---|
| 1 | 2 | 3 | 4 |
| 8·0 | 8·3 | 8·5 | 7·3 |
| 7·5 | 6·8 | 8·3 | 7·2 |
| 8·2 | 7·2 | 7·9 | 6·8 |
| 7·5 | 6·7 | 8·2 | 6·7 |
| 7·3 | | 8·4 | |

Do these data suggest that incisor width may vary significantly between these groups? To make life easier, only a few of the many hundreds of observations which would be needed in real life are listed.

# 9

# Measuring diagnostic consistency

Statistical analysis can only be as reliable as the data analysed. In clinical or epidemiological dental research, variations in the interpretation of diagnostic criteria can have a marked effect on the reliability of subsequent analysis. Methods of measuring examiner consistency in these circumstances are discussed, and illustrative examples provided.

All of the statistical techniques described in this book so far have been based on the assumption that the data being analysed are reliable. This seems, on the face of it, to be a statement too obvious to need making. Dental research, however, often depends on the way a researcher interprets diagnostic criteria, and on how his or her interpretation compares with someone else's. If, then, it appears that one examiner finds that children in Area A have a mean DMF of 3·5, while another examiner finds that in Area B children have a mean DMF of 3·0, we have to know whether the apparently lower DMF in Area B is due to a real difference in caries levels, is the result of sampling error, or is due to the fact that the two examiners do not agree on what constitutes a carious lesion.

In the early days of dental epidemiology, one reason for adopting a recognised diagnostic index (such as the DMF index) was because it was hoped that by standardising the diagnostic criteria, all examiners using that index would report the disease consistently. However, it soon became apparent that even the most detailed diagnostic criteria provided no guarantee of examiner consistency, and that even a single examiner could be inconsistent from one occasion to the next.

The problem has still to be satisfactorily resolved, and for the moment we have to accept that the comparison of data involving subjective clinical diagnoses by independent examiners can be unreliable. Even when examiners are carefully trained and calibrated against each other in the same survey, there is no guarantee that the results will be comparable, since aptitude, as well as training and experience, are involved here. One of many examples of the magnitude of this problem is provided in a recent paper by Nuttall and Davies,[1] who, in a follow-up to the 1983 Childrens' Dental Health Survey in Scotland, showed that 'only 51% of surfaces that were classified as decayed were subsequently treated on being presented for dental care within the following year', and that 'only 24% of the surfaces that were subsequently treated had been identified as decayed at the time of the survey' (excluding sound surfaces later extracted for orthodontic reasons or proximity to carious surfaces in the same tooth).

It is therefore vitally important that any survey involving the comparison of disease levels in two or more population groups must provide satisfactory factual evidence of the degree of diagnostic consistency and reliability of the examiner or examiners taking part (or, for that matter, of the degree of veracity or consistency in respondents to a questionnaire survey). The fact that most published reports falling into this category merely pay lip-service to the concept of examiner reliability by making such vague statements as 'Examiner consistency was tested and found to be within acceptable limits', or by reporting quite inadequate or useless tests of consistency, merely underlines the problems which still need to be faced.

At the moment, there appear to be no generally approved methods of measuring examiner variability. For example, both the World Health Organization[2] and the Fédération Dentaire Internationale[3] acknowledge the existence of the problem, and have suggested different solutions, but only in very general terms. As a result, it is possible to find, even in recently published papers, authors misapplying techniques which are perfectly acceptable in other contexts. Thus, simple $t$-tests may be used to assess examiner variability, even though these will only take into account whole mouth scores and ignore individual tooth or surface disagreements. As an extreme example, consider two examiners looking at the same group of subjects. Both report that 50% of all teeth examined are carious. Apparently, there is perfect agreement, but if, due to diagnostic confusion, half the teeth recorded as carious by one examiner were recorded as sound by the other and vice versa, there would still be 100% comparability in the overall results, but 0% examiner agreement! This means that in any subsequent survey involving these examiners, there would be no chance of diagnostic agreement! Correlation coefficients are also often used for this purpose, although these merely measure the strength of a relationship between two variables, not the agreement between them. For perfect agreement, the points have to lie along the line of equality (ie $y = x$), whereas for perfect correlation they may lie along any line. Bland and Altman[4] have made this point very forcefully, describing the correlation technique for testing reproducibility as 'totally inappropriate'.

The techniques to be described in the rest of this chapter are therefore offered as reasonably simple and practical working solutions to the problem, making use of statistical theory which has been covered earlier in this book. There are more sophisticated techniques available, but they require an understanding of complex statistics, with all the pitfalls that this implies for amateur statisticians.

## Dental caries—Present or absent

Two methods are described here of putting a numerical value on to inter- or intra-examiner variability, where either two examiners are involved or one examiner is used on two different occasions; the diagnosis is made on a 'yes/no', 'present/absent' or 'positive/negative' basis. It is a simple matter to devise a computer program (using a spreadsheet, for example) to carry out the necessary calculations. The descriptions conform to those contained in an excellent paper on the subject by Nuttall and Paul.[5]

### Data input

Data should be recorded as components of a $2 \times 2$ table, the four cells of which contain the following information:

(1) proportion of teeth both examiners agree are *sound* (a);
(2) proportion of teeth Examiner I considers to be *sound* but Examiner II considers to be *carious* (b);
(3) proportion of teeth Examiner I considers to be *carious*, but Examiner II considers to be *sound* (c);
(4) proportion of teeth both examiners agree are *carious* (d).

(Note that the *proportions*, and not the actual values are used in this analysis.) Thus, the table will look like this:

|  |  | Examiner I | | |
|---|---|---|---|---|
|  |  | Sound | Carious | Total |
|  | Sound | a | c | a+c |
| Examiner II | Carious | b | d | b+d |
|  | Total | a+b | c+d | a+b+c+d |

### Dice's Coincidence Index[6]

This index provides a measure of either the probability that a tooth (or surface) diagnosed as sound by one examiner will be diagnosed similarly by the other, or the probability that a tooth (or surface) diagnosed as carious by one examiner will be diagnosed similarly by the other.

The formula for calculating the first of these alternatives is

$$\frac{a}{[(a+b)+(a+c)]/2}$$

and the formula for calculating the second is

$$\frac{d}{[(c+d)+(b+d)]/2}$$

Thus, supposing that (in a caries calibration exercise, for example) we obtained the following data:

$$a = 0 \cdot 33, \quad b = 0 \cdot 05, \quad c = 0 \cdot 24, \quad d = 0 \cdot 37.$$

The probability of agreement for teeth diagnosed as sound would then be

$$\frac{0 \cdot 33}{[(0 \cdot 33 + 0 \cdot 05) + (0 \cdot 33 + 0 \cdot 24)]/2} = 0 \cdot 69 \text{ (or 69\%)}$$

The probability of agreement for teeth diagnosed as carious would be

$$\frac{0 \cdot 37}{[(0 \cdot 24 + 0 \cdot 37) + (0 \cdot 05 + 0 \cdot 37)]/2} = 0 \cdot 72 \text{ (or 72\%)}$$

Thus, in this case there is not much difference in the likelihood of agreement based on 'sound' diagnosis and agreement based on 'carious' diagnosis, although the latter is a shade more reliable.

### Cohen's kappa[7]

The kappa statistic relates the actual measure of agreement obtained with the degree of agreement which would have been attained had the diagnoses been made at random, or, in other words, the extent to which the actual degree of agreement recorded improves upon chance. This is probably the most reliable way of assessing overall examiner agreement.

The general formula is:

$$\frac{p_0 - p_e}{1 - p_e}$$

where $p_0$ is the proportion of observed agreement (ie $a+d$), and $p_e$ is the proportion of agreement which could be expected by chance. In our example, the proportion of agreement on 'sound' teeth to be expected by chance is $(a+c) \times (a+b)$, and the proportion of 'carious' teeth to be expected by chance is $(b+d) \times (c+d)$. If at the rate of diagnosis observed, the examiners had assigned teeth to the two categories at random, they could be expected to agree with each other on $[(a+c) \times (a+b)]+[(b+d) \times (c+d)]$ occasions overall. Thus, for our example:

$$p_0 = a+d = 0\cdot33+0\cdot37 = 0\cdot70$$
$$p_e = [(a+c)(a+b)]+[(b+d)(c+d)]$$
$$(0\cdot57\times0\cdot38)+(0\cdot42\times0\cdot61)$$
$$= 0\cdot47$$

So, kappa $= (0\cdot70 - 0\cdot47)/(1 - 0\cdot47) = 0\cdot23/0\cdot53$
$$= 0\cdot43$$

Incidentally, astute readers will already have spotted the similarity between this procedure and that for the calculation of chi-squared, which was described in an earlier chapter.

If the result is unity, then perfect agreement has been achieved. A negative score indicates that the examiners would have done better to have left things to chance, since they were diagnosing to different criteria. A kappa score of zero suggests that the examiners were classifying the teeth as if at random. It is suggested[8] that a score of over $0\cdot8$ indicates good agreement, over $0\cdot6$ indicates substantial agreement and over $0\cdot4$ moderate agreement. Our example score of $0\cdot43$, therefore, although showing 'moderate' agreement, is not very encouraging as the basis upon which to launch a survey!

## Periodontal disease and severity gradings

The previous section described methods of quantifying examiner variability in caries studies. There, since we are most frequently dealing with caries being recorded as either present or absent, the calculations are not too complex. However, if we were to try the same techniques for studies in which caries severity was measured, or in periodontal studies where most indices routinely use four grades of severity, we could find ourselves in difficulties.

It is, however, possible to modify the Cohen's kappa technique to fit these circumstances, and, indeed, a considerable amount of information on examiner performance can be obtained in the process. The following examples make use of the techniques described by Kingman.[9] They are based on the usual periodontal measurement scale of four grades: 0, 1, 2, 3 (or negative, mild, moderate, severe), but will work equally well with indices using a larger number of grades.

As before, the kappa formula is

$$k = (p_o - p_e)/(1 - p_e)$$

## Unweighted kappa

Suppose that after two examinations of the same group of 162 subjects (or units), the following contingency table was obtained (for, say, plaque index):

| Observed | | Exam I | | | | |
|---|---|---|---|---|---|---|
| | | 0 | 1 | 2 | 3 | Total |
| | 0 | 25 (0·154) | 22 (0·136) | 0 | 0 | 47 (0·290) |
| | 1 | 4 (0·025) | 54 (0·333) | 18 (0·111) | 0 | 76 (0·469) |
| Exam II | 2 | 0 | 10 (0·062) | 23 (0·142) | 3 (0·019) | 36 (0·222) |
| | 3 | 0 | 0 | 0 | 3 (0·019) | 3 (0·019) |
| Total | | 29 (0·179) | 86 (0·531) | 41 (0·253) | 6 (0·037) | 162 (1·00) |

In other words, 25 units were scored as '0' at both the first and second examinations, 22 were scored '1' at the first exam but '0' at the second, 4 were scored '0' at the first exam but '1' at the second, and so on. These raw data must now be converted into proportions of the total number of units examined (162); these proportions are shown in brackets in the Table. Thus, 25 as a proportion of 162 is 25/162 = 0·154.

The next step is to calculate 'expected' values (ie the values which would be expected if the scoring were purely random), which again follows the procedure used in calculating expected values in the chi-squared test. The top left-hand 'cell' will be: 47 (the top row total) × 29 (the left-hand column total)/162 (the grand total) = 8·41. These values must again be converted into proportions as before:

| Expected | | Exam I | | | | |
|---|---|---|---|---|---|---|
| | | 0 | 1 | 2 | 3 | Total |
| | 0 | 8·41 (0·052) | 24·95 (0·154) | 11·90 (0·073) | 1·74 (0·011) | 47 |
| | 1 | 13·60 (0·084) | 40·35 (0·249) | 19·23 (0·119) | 2·81 (0·017) | 76 |
| Exam II | 2 | 6·44 (0·040) | 19·11 (0·118) | 9·11 (0·056) | 1·33 (0·008) | 36 |
| | 3 | 0·54 (0·003) | 1·59 (0·010) | 0·76 (0·005) | 0·11 (0·001) | 3 |
| Total | | 29 | 86 | 41 | 6 | 162 |

Expected value = (row total × column total)/grand total

Next, $p_o$, the *observed* probability of agreement is calculated. This is the sum of the *observed* proportions in cells '0,0'; '1,1'; '2,2'; and '3,3'. That is

$$0 \cdot 154 + 0 \cdot 333 + 0 \cdot 142 + 0 \cdot 019 = 0 \cdot 648$$

Finally, $p_e$, the *expected* probability of agreement is found in the same way. In this case it is

$$0 \cdot 052 + 0 \cdot 249 + 0 \cdot 056 + 0 \cdot 001 = 0 \cdot 358$$

Therefore, $p_o - p_e = 0 \cdot 648 - 0 \cdot 358 = 0 \cdot 290$

and $\qquad 1 - p_e = 1 \cdot 00 - 0 \cdot 358 = 0 \cdot 642$

Kappa is then $0 \cdot 290 / 0 \cdot 262 = 0 \cdot 45$ (ie barely 'moderate' agreement).

### Weighted kappa

Unweighted kappa only considers areas of *total* agreement; it takes no account of 'near-misses'. But a 'near-miss', although not as good as total agreement, is still better than total disagreement. Therefore some sort of weighting system must be applied to such scores, so that they can make an appropriate and realistic contribution to the kappa statistic.

In this context, the somewhat arbitrary weighting system devised by Cicchetti[10] has the advantage that it 'penalises' errors in disease/no disease gradings more severely than the less critical errors in severity gradings:

Restricted linear weights

|  |  | Exam I |  |  |  |
|---|---|---|---|---|---|
|  | 0 | 1 | 2 | 3 | 4 |
| 0 | 1·00 | 0·60 | 0·20 | 0·00 | 0·00 |
| 1 | 0·60 | 1·00 | 0·80 | 0·40 | 0·00 |
| Exam II   2 | 0·20 | 0·80 | 1·00 | 0·80 | 0·40 |
| 3 | 0:00 | 0·40 | 0·80 | 1·00 | 0·80 |
| 4 | 0·00 | 0·00 | 0·40 | 0·80 | 1·00 |

Note that these weightings are shown for a five-grade index, to give a better idea of how they are derived.

They are applied by multiplying each proportion in the 'observed' and 'expected' contingency tables by the appropriate weight. The proportion in cell 0,0 will therefore be multiplied by $1 \cdot 00$; 0,1 will be multiplied by $0 \cdot 60$, and so on. Thus, the greatest contribution to the total score is made by cells where agreement was highest (ie 0,0; 1,1; 2,2), and the least contribution is made by cells where the agreement was low (ie 0,2; 1,3; 3,1).

This will give us the following results

## Weighted observed p

|  |  | Exam I | | | |
|---|---|---|---|---|---|
|  |  | 0 | 1 | 2 | 3 |
|  | 0 | $0 \cdot 154$ | $0 \cdot 082$ | 0 | 0 |
| Exam II | 1 | $0 \cdot 015$ | $0 \cdot 333$ | $0 \cdot 089$ | 0 |
|  | 2 | 0 | $0 \cdot 050$ | $0 \cdot 142$ | $0 \cdot 015$ |
|  | 3 | 0 | 0 | 0 | $0 \cdot 019$ |

## Weighted expected p

|  |  | Exam I | | | |
|---|---|---|---|---|---|
|  |  | 0 | 1 | 2 | 3 |
|  | 0 | $0 \cdot 052$ | $0 \cdot 092$ | $0 \cdot 015$ | 0 |
| Exam II | 1 | $0 \cdot 050$ | $0 \cdot 249$ | $0 \cdot 095$ | $0 \cdot 007$ |
|  | 2 | $0 \cdot 008$ | $0 \cdot 094$ | $0 \cdot 056$ | $0 \cdot 006$ |
|  | 3 | 0 | $0 \cdot 004$ | $0 \cdot 004$ | $0 \cdot 001$ |

$p_0 = 0 \cdot 899$ (this is now the total of all the weighted values)

$p_e = 0 \cdot 733$   $p_0 - p_e = 0 \cdot 166$   $1 - p_e = 0 \cdot 267$   Wtd $k = 0 \cdot 62$

This raises the status to marginally 'substantial agreement'.

**Discriminatory weightings**

One of the big advantages of this kappa analysis is that it becomes possible, by the use of discriminatory weightings on the original data, to analyse where the weakest areas of diagnosis lie. For example, is most confusion being caused by interpretation of the '0' grading, or the '1' grading, or even the difference between '0' and '1'? Would markedly greater examiner consistency be achieved if the lower two grades and the upper two grades were combined to produce a two-level instead of a four-level index? Space does not permit a full discussion of these refinements, but a detailed description may be found in the paper by Kingman already referred to.[9]

**Quantitative data**

This section summarises the approach suggested by Bland and Altman[4] as a more appropriate alternative to using a correlation coefficient, when quantified scores are being compared. First of all, it is useful routinely to prepare a scatter diagram for the data, when the results should approximate to a straight line (hence suggesting the correlation approach). However, a more realistic association is revealed by plotting the differences between the individual pairs of readings (y) against their average (x).

At this point it is probably better to illustrate the technique with a worked example. The data in Table I represent the cumulative pathological pocket depths in millimetres in seven periodontal patients, as measured by two independent examiners, A and B. If the paired readings for each subject are now plotted in a scatter diagram, the result will look something like figure 1. From this it can be seen that the points approximate very closely

**Table I**  Comparison of cumulative pocket depth (mm) in seven patients as recorded by two independent examiners

| Subject | Exam A | Exam B | B − A | Mean, A, B |
|---------|--------|--------|-------|------------|
| 1 | 14·7 | 15·6 | 0·90 | 15·15 |
| 2 | 9·8 | 11·5 | 1·70 | 10·65 |
| 3 | 15·8 | 16·0 | 0·20 | 15·90 |
| 4 | 11·7 | 11·4 | −0·30 | 11·55 |
| 5 | 13·8 | 15·0 | 1·20 | 14·40 |
| 6 | 17·9 | 20·0 | 2·10 | 18·95 |
| 7 | 12·1 | 10·0 | −2·10 | 11·05 |

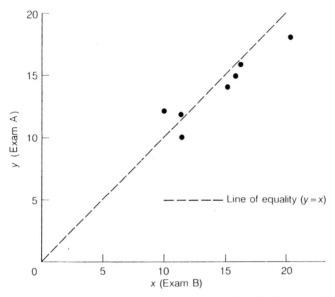

**Fig. 1** Scatter diagram of cumulative gingival pocket depths as estimated by two independent examiners (data from Table I).

to the line of equality ($y = x$). It is not surprising to find that the correlation coefficient for the association between $x$ and $y$ is $0.92$, indicating a high degree of correlation. However, this does not imply a high degree of examiner consistency, as subsequent analysis will show.

Figure 2 provides a plot of the differences between the two readings for each subject against the mean of the two readings. This diagram also shows the overall mean of all the individual differences A − B ($0.529$) together with the mean ± 2SDs (where $s = 1.4221$). The lower limit is $-2.315$ and the upper is $3.373$. In other words, we would usually expect examiner disagreement (B − A) to be between B scoring $3.4$ mm more than A, and A scoring $2.3$ mm more than B. Since the lowest actual score is of the order of 10, this represents a maximum possible error of the order of 30%, a very different result from the near-perfect correlation coefficient.

Furthermore, this result is obtained from only one sample. Calculation of the standard error of the mean will indicate the likely range of differences between A and B if several similar samples were taken. The formulae used here are

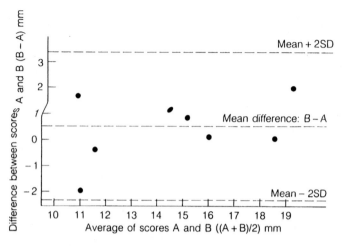

**Fig. 2** Plot of the difference in cumulative gingival pocket depths, as measured by two independent examiners, against the mean of the two measurements (data from Table I).

$$\text{SE } (d) = \sqrt{s^2/n} = \sqrt{1 \cdot 4221^2/7} = 0 \cdot 5375$$

$$\text{SE limit values} = \sqrt{3s^2/n} = \sqrt{6 \cdot 0671/7} = 0 \cdot 9310$$

Thus, 95% CI for the true mean difference B − A is

$$0 \cdot 529 \pm t(0 \cdot 05, 6) \times 0 \cdot 5375$$
$$= -0 \cdot 79 \text{ to } 1 \cdot 85$$

This range is particularly useful because it indicates that although the mean (0·529) shows that B tends to measure higher than A, this apparent bias is within the limits to be expected from sampling variation.

The 95% CI for the upper limit is

$$3 \cdot 373 \pm t(0 \cdot 05, 6) \times 0 \cdot 9310$$
$$= 1 \cdot 09 \text{ to } 5 \cdot 65$$

The 95% CI for the lower limit is

$$-2 \cdot 315 \pm t(0 \cdot 05, 6) \times 0 \cdot 9310$$
$$= -4 \cdot 50 \text{ to } -0 \cdot 34$$

These figures show very clearly that although the average difference between the two examiners may be quite small, the range of likely disagreement on individual subjects may be remarkably high. The importance of this is not that it may 'discredit' a significant finding in subsequent statistical analysis, but rather that it may render a truly significant difference impossible to demonstrate. This is because each of the two (or more) samples under review will inevitably incorporate these random examiner errors which will, in turn, affect the standard deviation of any true differences present. An otherwise identical comparison of the difference between two sample means could therefore prove to be significant if the examiners showed no variability, but not significant if the examiners were inconsistent. The greater the examiner variability, the more difficult it becomes to demonstrate a true difference statistically.

For example, suppose the 'Examiner A' data ($x = 13 \cdot 7$, $s = 2 \cdot 73$) were to be compared with data obtained from another group of patients with no examiner variability involved. Suppose the mean for this group is $10 \cdot 1$, with sample size and standard deviation as before.

Then

$$t = (13 \cdot 7 - 10 \cdot 1)/1 \cdot 461 = 2 \cdot 46 \quad (P = 0 \cdot 05)$$

In other words, a difference in means of only $3 \cdot 6$ mm is significant at the 5% level. However, in practice, as we have seen, examiner variability could change this value, at the most conservative estimate, to anything between $2 \cdot 3$ and $4 \cdot 9$, thus effectively 'masking' the true difference between the sample means. This method is, of course, not appropriate for caries or periodontal consistency analysis, for reasons already given above.

## References

1 Nuttall N M, Davies Janet A. The capability of the 1983 Children's Dental Health Survey in Scotland to predict fillings and extractions subsequently undertaken. *Community Dent Health* 1988; **5**: 355–362.

2 World Health Organization. *Oral health surveys. Basic methods*. 3rd ed. Geneva: WHO, 1987.

3 Fédération Dentaire Internationale. Technical Report No. 1. Principal requirements for controlled clinical trials of caries preventive agents and procedures. 3rd ed. London: FDI, 1982.

4 Bland J M, Altman D G. Statistical methods for assessing agreement between two methods of clinical measurement. *Lancet* 1986; **1**: 307–310.

5 Nuttall N M, Paul J W McK. The analysis of inter-dentist agreement in caries prevalence studies. *Community Dent Health* 1985; **2**: 123–128.

6 Dice L R. Measures of the amount of ecological association between species. *Ecology* 1945; **26:** 297–302.
7 Cohen J. A coefficient of agreement for nominal scales. *Educ Psychol Measurement* 1960; **20:** 37–46.
8 Landis J R, Koch G G. The measurement of observer agreement for categorical data. *Biometrics* 1977; **33:** 159–174.
9 Kingman A. A procedure for evaluating the reliability of a gingivitis index. *J Clin Periodontol* 1986; **13:** 385–391.
10 Cicchetti D V. Assessing inter-rated reliability for rating scales; resolving some basic issues. *Br J Psychiatry* 1976; **129:** 452–456.

## Practical example

A dental survey researcher looks at a random sample of 200 deciduous teeth on two separate occasions. Of the 200, he records 130 as being 'sound' on both occasions and 54 as being 'carious' on both occasions. Of the remaining 16, he finds 10 'carious' on the first occasion but 'sound' on the second occasion, and 6 'sound' on the first occasion but 'carious' on the second.

Calculate a reproducibility coefficient (Dice's Index) for both 'carious' and 'sound' diagnoses.

Suppose this same examiner now carried out a survey in which he reported that the mean dmft was $1 \cdot 33$ in a sample of 500 5-year-old children (dt = $0 \cdot 95$, mt = $0 \cdot 15$, ft = $0 \cdot 24$). What effect would his reproducibility coefficient be likely to have on the result, were he to repeat the survey? Assume that all children have a normal deciduous dentition (no permanent teeth).

*Note:* Although when reporting the results of such a survey, the basic experimental unit must be the individual rather than the tooth, in order to ensure that all observations are independent, this proviso does not apply perhaps so importantly when assessing diagnostic consistency, since in this case the experimental unit is the unit upon which the diagnosis is made; ie the tooth itself.

# 10

# Life tables and survival analysis

Methods based on the construction of life tables are described which may be used to estimate the likely survival time of individual tooth types, restorations or whole dentitions, and to compare general oral health levels in different populations or in the same population at different times.

## Life tables

The life-table technique described here is more commonly used to answer demographic questions such as: 'Suppose that in a population a convenient number, say 1000 or 10 000 babies, were born on the same day; how many of the babies would survive to celebrate their first, second, third . . . or in general $x$th birthday, assuming that the babies died at the *current* rate of mortality?'

It requires little imagination to see how this may be adapted to the dental context; we merely substitute teeth erupted or fillings placed for babies born. However, this type of question, whether demographic or dental, is unreal, mainly because the 1000 or 10 000 (an arbitrary figure known as the *radix*) babies would, in fact, die at the rates existing at the time *when* they die, and not at the rates existing when they were born. The life table is therefore a fiction, but it is a fiction which describes current mortality levels at each age, and can therefore be used to make comparisons between different population groups or between the same groups at different points in time. A full life table includes every single year of existence from 0 to the highest age of survival, while an abridged life table considers only 5-yearly age groupings, although greater detail may be included in the first 5-year period.

Table I shows a typical abridged life table for dentition survival in an English adult community in the 1960s.[1] Here the radix is 1000, that is, the table describes the dentition mortality of 1000 subjects, assuming they become edentulous at the mortality rates which existed in that community as a whole in 1965, when the survey on which the table is based was carried out.

91

**Table I**   Abridged mortality table: Total tooth loss in an English population, 1965

| $Age_x$ | % Edentulous | $l_x$ | $d_x$ | $p_x$ | $q_x$ | $e_x^0$ |
|---|---|---|---|---|---|---|
| | | 1000 | 13 | 0·987 | 0·013 | 37·25 |
| 15 | 1·3 | 987 | 17 | 0·983 | 0·017 | 32·71 |
| 20 | 3·0 | 970 | 32 | 0·967 | 0·033 | 28·24 |
| 25 | 6·2 | 938 | 54 | 0·942 | 0·058 | 24·12 |
| 30 | 11·6 | 884 | 83 | 0·906 | 0·094 | 20·44 |
| 35 | 19·9 | 801 | 110 | 0·863 | 0·137 | 17·30 |
| 40 | 30·9 | 691 | 129 | 0·813 | 0·187 | 14·66 |
| 45 | 43·8 | 562 | 138 | 0·755 | 0·245 | 12·45 |
| 50 | 57·6 | 424 | 129 | 0·696 | 0·304 | 10·68 |
| 55 | 70·5 | 295 | 106 | 0·641 | 0·359 | 9·26 |
| 60 | 81·1 | 189 | 80 | 0·577 | 0·423 | 8·06 |
| 65 | 89·1 | 109 | 51 | 0·532 | 0·468 | 7·13 |
| 70 | 94·2 | 58 | 31 | 0·466 | 0·534 | 6·21 |
| 75 | 97·3 | 27 | 15 | 0·445 | 0·555 | 5·46 |
| 80 | 98·8 | 12 | 8 | 0·333 | 0·667 | 4·17 |
| 85 | 99·6 | 4 | 4 | 0·000 | 1·000 | — |

The column headings may be described as follows:

$Age_x$ is the age in 5-year intervals to which the numbers in the other columns relate.

$l_x$ gives the number of subjects still retaining some natural teeth at exact age $x$. Thus, 970 are dentate at age 20, and 12 are still dentate at age 80.

$d_x$ represents the number of subjects becoming edentulous between exact age $x$ and exact age $x+5$. Thus, the number becoming edentulous between 30 and 35 is 83. In general, $d_x = l_x - l_{x+5}$.

$p_x$ is the probability of remaining dentate from exact age $x$ to exact age $x+5$. Thus, the probability that a subject who is dentate on his 60th birthday remains dentate to his 65th birthday is 0·577. This is $l_{x+5}/l_x$, which in this case is $109/189 = 0·57672$.

$q_x$ is the probability of becoming edentulous between exact age $x$ and exact age $x+5$. For any given year, $p_x + q_x$ must equal 1. Thus, $q_x = 1 - p_x = d_x/l_x$.

$e_x^0$ is the 'life expectancy'; in this case, the average number of years to be lived from age $x$ before becoming edentulous. Note the *average* number of years; some individuals may become edentulous well before $e_x^0$, while others may remain dentate for some time afterwards. $e_x^0$ is calculated by summating $l_{x+1}, l_{x+2}, \ldots l_{x+n}$, multiplying the total by the time interval, dividing the new total by $l_x$ and adding half the time interval to the result. Thus $e_x^0$ for age 30 is:

$$[(5(801 + 691 + 562 + 424 + \ldots + 12 + 4))/884] + 2 \cdot 5 = 20 \cdot 44 \text{ years.}$$

This calculation is only approximate, since it is assumed that subjects become edentulous evenly throughout each 5-year period. This may or may not be true for any given population.

Life tables were first constructed and used by actuaries for the purpose of calculating premiums to be paid for life insurance, but their usefulness in the medical field soon became apparent. In dentistry their potential remains virtually untapped. Since the data required to construct a dentition life table is merely the number of teeth present per subject, together with the subject's age, the method is not distorted by examiner variability (providing the examiner can count up to 32!), and therefore it provides a valuable method of comparing general oral health in different countries, just as life expectancy data are used in the medical field. Furthermore, the method is sensitive enough to register quite small changes in tooth loss patterns, so that the procedure provides a useful method of monitoring the success of dental care systems or treatment philosophies.

In medical research, the technique has perhaps its most useful application in follow-up or survivorship studies. For example, in investigations of the distribution of survival times after some critical event, such as the appearance of symptoms, admission to hospital, or completion of a prescribed form of treatment. In dentistry this would include the life of restorations in general, or by particular type, or the survival of teeth which had been root-filled, for example. Once again, a life table is constructed, but the $x$ column, rather than measuring the age of the subjects, indicates the length of time since the critical event. The $l_x$ column states the numbers of survivors (from a suitable radix), $x$ weeks or months or years after the critical event.

## Survival analysis

In a life table, given survival to age $x$, the probability of dying between age $x$ and $x+n$ is:

$$(l_x - l_{x+n})/l_x$$

Now this will be small for small values of $n$, but will get larger as $n$ increases. In fact, we could write the probability of dying between age $x$ and $x+n$ as:

$$[(l_{x - 1x + n})/nl_x]n$$

Thus, for values of $n$ which are not too large, this probability is proportional to $n$, since the part of the equation in the square brackets is approximately constant. If we consider smaller and smaller values of $n$, the limit of this constant, as $n$ tends to zero, is called the force of mortality, the failure rate, or the hazard at age $x$, and this is denoted by $\mu_x$.

In the sort of survivorship analysis we are interested in, we are replacing age, $x$, by survival time, $t$. Of particular interest is the situation where $\mu_t$ is constant, that is, independent of time $t$. In this situation, we do not need to specify $t$ in $\mu_t$. Let us call this constant force of mortality $\lambda$.

If the annual probability of 'death' (or failure), $q$, is constant, then:

Pr ('death' in 1st year) $= q$
Pr ('death' in 2nd year) $= (1 - q)q$
Pr ('death' in 3rd year) $= (1 - q)^2q$
Pr ('death' in year t) $= (1 - q)^{t-1}q$ and
Pr ('death' any time after $t$) $= (1 - q)^t$

In this situation, the successive probabilities make up a *geometric series*, and $t$ is said to follow a geometric distribution. It is not too difficult to show that the mean survival time is $1/q$ and the variance of survival times is $(1 - q)/q^2$.

We have regarded survival time as a discrete variable; in the above, $t$ has taken values 1, 2, 3 . . . years. In reality, of course, $t$ is continuous and although the geometric distribution may provide an approximate description of survivorship, particularly when $q$ is small, a better description of survivorship would be obtained if $t$ were regarded as a continuous variable. Now, if instead of working with $t$ in years, suppose we divide each year into $k$ equal intervals (if $k = 12$, we would be measuring time in months; if $k = 52$, we would have weeks, and so on). What happens to Pr ('death' any time after $t$)? This is, of course, Pr (survival time greater than $t$) and is

$$(1 - q/k)^{kt}$$

What happens to this if $k$ gets larger and larger, ie as discrete time gets closer and closer to continuous time? It can be shown, using the binomial theorem, that in continuous time

$$\text{Pr (survival greater than } t) = e^{-\lambda t}$$

and $t$ is said to follow an exponential distribution. The mean survival time is $1/\lambda$ and the median survival time $t_m$ is $0\cdot693/\lambda$, because if

$$e^{-\lambda t_m} = 0\cdot5, \quad -\lambda t_m = \log_e 0\cdot5 = -0\cdot693$$

$$\text{and so } t_m = 0\cdot693/\lambda$$

If the force of mortality $\mu_x$ really is constant and equal to $\lambda$, then Pr (survival time greater than $t$) when plotted on ordinary graph paper, should look like figure 1, while since $\log_e (e^{-\lambda t}) = -\lambda t$, if $\log_e (e^{-\lambda t})$ is plotted vertically (equivalently if $e^{-\lambda t}$ is plotted on log scale) the result will be a straight line with slope $-\lambda$ (fig. 2). This is why survival curves are often drawn on semi-log graph paper.

## Construction of survivorship tables
If it can be assumed that $\mu_x$ is constant and equal to $\lambda$, then there is no problem, since the probability of surviving more than $t$ years (or whatever time units) is $e^{-\lambda t}$ and this can be calculated directly. However, it is not

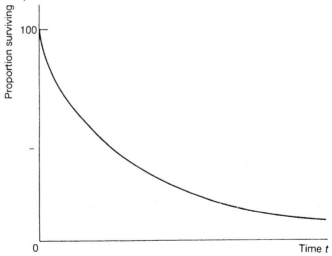

**Fig. 1** Pr (survival time greater than $t$) if the force of mortality $\mu_x$ is constant and equal to $\lambda$. Plotted on ordinary arithmetic graph paper.

usually possible to assume constant $\mu_x$ without first constructing the survivorship table.

One further complication remains. If the results of survival times are based on a relatively short period of follow-up, some survival times may be *censored*. A censored survival time of say 5 years refers to a subject who was last seen 'alive' 5 years after time zero (the time of diagnosis, or the time of placement of a restoration, for example). The full survival time is therefore unknown, but it must be more than 5 years. Clearly, to omit patients with censored survival times from the analysis would be a serious bias. A method which takes account of incomplete follow-up is known as the Kaplan-Meier method, which is best illustrated by an example, showing how even in cases where the proportion of censored cases is high, useful information on the survivorship pattern may still be obtained.

In a study of the durability of amalgam restorations, 23 premolar teeth which had been filled on the buccal surface were followed up to determine when the restoration failed. The survival times are shown below in months, where figures with a '+' after them indicate that the filling was still satisfactory when last seen that number of months after insertion. Its fate after that point in time is unknown.

5, 10, 13, 16, 23, 24, 25, 26, 26, 31
19+, 28+, 28+, 34+, 35+, 40+, 40+, 46+, 46+, 46+, 49+, 52+

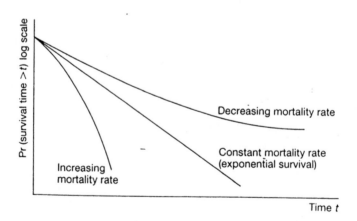

**Fig. 2** Pr (survival time greater than *t*) if $\mu_x$ is constant and equal to λ. Plotted on semi-log graph paper.

**Table II**  Calculation of Kaplan-Meier probabilities of survival for premolar restoration data

| $t$ | $n_t$ | $d_t$ | $n_t - d_t$ | $p_t$ | Pr(surv. $> t) = 1_t$ |
|-----|-------|-------|-------------|-------|----------------------|
| 0   | 23    | 0     | 23          | 1       | 1       |
| 5   | 23    | 1     | 22          | 0·9565  | 0·9565  |
| 10  | 22    | 1     | 21          | 0·9545  | 0·9130  |
| 13  | 21    | 1     | 20          | 0·9524  | 0·8696  |
| 16  | 20    | 1     | 19          | 0·9500  | 0·8261  |
| 23  | 18    | 1     | 17          | 0·9444  | 0·7802  |
| 24  | 17    | 1     | 16          | 0·9412  | 0·7343  |
| 25  | 16    | 1     | 15          | 0·9375  | 0·6884  |
| 26  | 15    | 2     | 13          | 0·8667  | 0·5966  |
| 31  | 11    | 1     | 10          | 0·9091  | 0·5424  |

Table II shows the calculation of the Kaplan-Meier probabilities of survival. In this table:

$t$ shows the times at which the fillings failed.

$n_t$ is the number of fillings exposed to risk of failure at time $t$.

$d_t$ is the number of failures observed at time $t$.

$p_t$ is the estimated probability of survival between successive time entries in the table.

$1_t$ is the product of all the earlier values of $p_t$.

Note that:

(1) $1_t$ is only estimated for those values of $t$ which are complete survival times.

(2) $n_t$ decreases as $t$ increases because of failures and losses due to censored follow-up. Thus, for $t = 23$, $n_{23} = n_{16} - 1$ (failure) $-1$ (censored observation 19+ between $t = 16$ and $t = 23$).

If $1_t$ is plotted against $t$ on semi-log graph paper it is seen (fig. 3) that over the first part of the survivorship curve the trend is approximately linear, implying at least approximately an exponential distribution of survival times.

The estimated failure rate $\lambda$ is

No. of failures/Total follow-up time

ie est. $\lambda = 10/(5 + 10 + \ldots + 31 + 19 + 28 + \ldots + 52)$

$= 10/708 = 0·014$ per month.

An exponential survival curve (straight line) is plotted for comparison on the same graph and seems to indicate, if anything with such small numbers, that $\mu_t$ may not be constant but increasing as $t$ increases. Notice that use of the exponential model enables an estimate of the mean and median survival times to be made.

Since $\mu = 1/\lambda$, $\bar{x} = 1/0 \cdot 014 = 70 \cdot 8$ months

and the estimated median $x = 70 \cdot 8 \log_e 0 \cdot 5 = 49 \cdot 07$ months.

These estimates could not be made using conventional life table techniques, but they may be very imprecise, particularly if the final proportion surviving is still large, as it is in this example. A good general discussion of survival studies may be found in two papers by Peto *et al.*[2,3]

## Comparison of two exponential survival curves

If it became of interest to compare, for instance, the survival curves for amalgam versus gold restorations in a given type of cavity, the first step would be to construct Kaplan-Meier survivorship curves for each material, in order to check for major departure from the exponential model. These are therefore plotted on semi-logarithmic graph paper and should each approximate to a straight line. If so, the estimated 'mortality' rates ($\lambda$) can be calculated for each material, as shown above. A simple test of the ratio, R, of these two rates is achieved by considering $\log_e R$. It can be shown that $\log_e \lambda$ is approximately Normally distributed, with variance $1/d$ where $d$ is the observed number of 'deaths'. Thus,

$$SND = \log_e R / \sqrt{(1/d_a + 1/d_g)}.$$

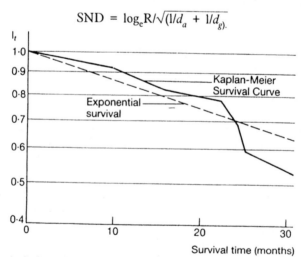

Survival time (months)

**Fig. 3** Survival of premolar restorations: Kaplan-Meier and exponential survival lines.

A 95% confidence interval for the logarithm of the ratio of the rates can then be calculated, which will not be symmetrical about the observed ratio. For the non-parametric comparisons of two survival curves, log-rank tests may be used, but these are quite complex and space does not permit their inclusion here. Interested readers may be referred to any good textbook of medical statistics, such as Armitage and Berry.[4] Caution should also be used in interpreting non-exponential survivorship curves, as these may change in relation to each other with time. Thus, median or mean survival times, or the overall mortality rate, will have little meaning.

## References

1 Bulman J S. An epidemiological study into the prevalence of dental disease in sample populations from two contrasting English towns and their present and future treatment needs. p152. London: University of London, 1971. PhD Thesis.

2 Peto R, Pike M, Armitage P, Breslow N E, Cox D R, Howard S V, Mantel N, McPherson K, Peto J, Smith P G. Design and analysis of randomized clinical trials requiring prolonged observation of each patient. I. Introduction and design. *Br J Cancer* 1976; **34**: 585.

3 Peto R, *et al*. Analysis and examples. *Br J Cancer* 1977; **35**: 1.

4 Armitage P, Berry G. *Statistical methods in medical research*. 2nd ed. Oxford: Blackwell, 1987.

# 11

# Statistical analysis — The future

We have come to the last of these presentations on statistics as applied to dental research, and it could be that these words are read with a sigh of relief and the consoling thought that perhaps there is not much more to data analysis than the contents of these eleven simple chapters. Sadly, there is no such comfort here. These chapters have merely highlighted some of the more common methods available to analyse dental data, and have been presented and illustrated with data sets that are much simpler than those commonly encountered in real life. For example, to compare the means of a quantitative response in two independent groups of patients, we first check the assumption of Normality and constant variance, and, if satisfied (at least approximately), we can then do a $t$-test and calculate a confidence interval for the true difference. Unfortunately, data are seldom as simple as this. There may be a host of other characteristics recorded for each patient, which may affect the response and yet be unequally distributed between the two groups. Does it matter, for example, that one group has more males than the other, or that one group had a slightly higher baseline figure than the other? Is it possible to compare the two groups, taking account of other variables that have also been recorded? If not, it would seem to have been a waste of time to have recorded them in the first place.

The answer to the last question is, of course, yes; the extra information can, and more importantly, should be used in the analysis. However, the techniques necessary are rather more arithmetically complicated than those used up to now, and, these days, would only be attempted with the aid of a computer and suitable software. In fact, these methods, which are known as 'multiple regression' or 'linear models', have been theoretically understood for decades, but are only now becoming widely used as a result of the improved availability of computers.

It may come as a surprise that the methods already described in this book, including $t$-tests and chi-squared tests, can all be handled using multiple regression, although for simple data sets multiple regression is much more cumbersome than, say, a straightforward $t$-test.

## Statistical models

A statistical model is an algebraic equation which relates one variable, $y$, to potential explanatory variables, which may be designated $x_1, x_2, x_3 \ldots$ We have described a simple example of this in the chapter on linear regression. There, the response variable ($y$) is, we presume, linearly related to a single $x$ variable in such a way that the equation

$$y = \alpha + \beta x$$

provides an overall description of the relationship between $y$ and $x$. Obviously, it does not describe the data perfectly; the data points may be quite widely scattered about the line represented by this equation. The equation merely provides a *model* to describe the relationship, and this model is fitted to the data in such a way that we can obtain $a$ and $b$, which are the best estimates of the parameters $\alpha$ and $\beta$. Thus, the equation calculated to describe the data is

$$y = a + bx$$

In fact, the simplest model to describe the response variable is just the arithmetic mean of $y$, but this would tell us nothing about the effect of $x$ on $y$. We fit $x$ into the model in order to investigate whether $x$ can explain some of the variability of $y$, and the significance of the effect of $x$ on $y$ can be tested (and is routinely tested by most computer programs) using a $t$-test. The effect of $x$ on $y$ is indicated by the value of $b$, because $b$ is the slope of the line and so, on average, the value of $y$ increases (or decreases) by $b$ for each unit increase in $x$. The null hypothesis being tested here is that $\beta = 0$, and the $t$ value is $(b-0)/\text{Est.SE}(b)$. Figure 1 shows diagrams of some simple models.

## Linear models for parametric analyses

Let us suppose that we have two treatments, A and B, which we wish to compare, and that $y$ is a continuous variable: the response to the treatment. If we randomly allocate our patients to the two treatment groups, we might test the significance of the difference between them by calculating

$$t = (\bar{y}_A - \bar{y}_B)/\text{Est.SE}(\bar{y}_A - \bar{y}_B)$$

It may seem surprising, but this problem can be expressed in terms of

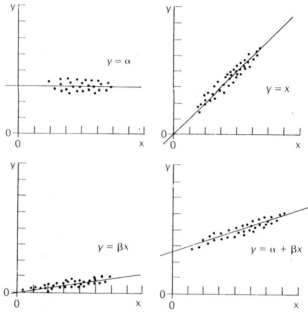

**Fig. 1**  Some simple statistical models.

linear regression by the use of a statistical trick. Suppose we define a variable $x$, which has the special property that $x = 1$ for an observation in group B and $x = 0$ for an observation from group A. A variable like this is often called a 'dummy' variable. In this situation, the scatter diagram of $y$ against $x$ would be similar to figure 2.

We could fit the model $y = \alpha + \beta x$ by calculating $a$ and $b$ so that the fitted line would be $y = a + bx$, but how would $a$ and $b$ be interpreted?

Suppose $x = 0$, that is, the observation is from group A. Then $y = a + b \times 0 = a$. So we have the very simple model for group A that $a$ must be equal to $\bar{y}_A$. However, what if $x = 1$, that is, an observation from group B? The equation now is $y = a + b \times 1$, and $a + b \times 1$ must equal $\bar{y}_B$.

Thus, $b = \bar{y}_B - \bar{y}_A$ and the test of significance of $b$ is exactly equivalent to the test of the difference between two sample means. In fact, all the parametric analyses given in basic textbooks can be expressed in terms of linear regression by specifying a linear model. If a good computer regression package is available, all the other tests and procedures can be carried out, although often with an increase in arithmetic computation.

The beauty of the linear model approach is that it can be extended,

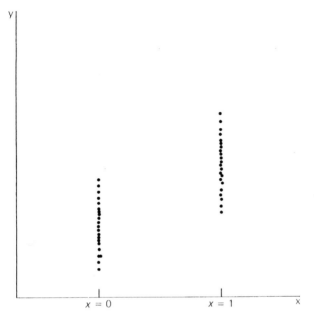

**Fig. 2**   Scatter diagram for a dummy variable defining two experimental groups.

using multiple regression, and used for the more complicated data sets, including one-way and two-way analysis of variance, even if the data set is unbalanced.

## More than one explanatory variable

Suppose we have the situation where a response variable is associated with two explanatory variables, $x_1$ and $x_2$, and suppose that (using a multiple regression computer program) we fit the model

$$y = \alpha + \beta_1 x_1 + \beta_2 x_2$$

The computer will provide us with the best estimates of $\alpha$, $\beta_1$ and $\beta_2$ and the fitted model will be

$$y = a + b_1 x_1 + b_2 x_2$$

How can the values of $b_1$ and $b_2$ be interpreted? Suppose we consider a situation where the response variable ($y$) is the durability (lifespan) of dental restorations, and the explanatory variables are the volume of the restoration ($x_1$) and the number of times it has been replaced ($x_2$). For restorations of a given volume, then, ($x_1$), $b_2$ is the effect of the number of replacements on the durability when the volume of the restoration is kept constant. That is, $b_2$ is the independent effect of the number of replacements on $y$. Similarly, $b_1$ is the effect of restoration volume for restorations which have been replaced a given number of times. Thus, although $x_1$ and $x_2$ are associated (restorations which have been replaced a number of times tend to be bigger), the linear model, including both explanatory variables, is able to separate their effects. Note that in this particular example, the $b$ values will almost certainly be negative, as there will be an inverse relationship between $y$ and both $x_1$ and $x_2$ (the bigger the restoration, and the greater the number of times it has been replaced, the shorter the life of the restoration).

Supposing that some of the restorations were gold and some were amalgam. We could define an explanatory variable $x_3$ to 'measure' restoration material, so that $x_3 = 0$ for amalgam and $x_3 = 1$ for gold, and fit the model

$$y = \alpha + \beta_1 x_1 + \beta_2 x_2 + \beta_3 x_3$$

We obtain the estimates of $\alpha$, $\beta_1$, $\beta_2$, and $\beta_3$ and the fitted equation is

$$y = a + b_1 x_1 + b_2 x_2 + b_3 x_3$$

Now, $b_1$ is the average decrease in $y$ per single unit of volume for restorations at a given replacement level and composed of the same material; $b_2$ is the average decrease in $y$ for restorations of a given volume and similar restoration material, and $b_3$ is the average difference between gold and amalgam for a given volume and replacement level. Thus, we have disentangled the effect of each factor by *adjusting* for the effect of the other two.

In this way, multiple regression can be used to analyse complex data sets, by defining the response variable $y$ and measuring the explanatory variables $x_1, x_2, x_3 \ldots$ where $x_1, x_2, x_3$ may be either continuous variables or dummy variables defining categorical variables (such as gender).

In order to use a computer program, the data will need to be arranged in a rectangular array, so that each line represents a patient (or, more generally, a 'record') and the values of $y$ and $x_1, x_2, x_3 \ldots$ form the columns of the array. Such an array is sometimes called a data matrix.

There are, of course, assumptions made in the preceding description of multiple regression analysis. In particular, it is assumed that there is a linear relationship between $y$ and each of the explanatory variables. Also, it is assumed that $y$ is Normally distributed with constant variance. Very often, however, $y$ is not Normally distributed, being skew to the right (positively skew), and neither is the variance of $y$ constant.

Commonly, the standard deviation increases as the magnitude of $y$ increases. In such circumstances, it may be preferable to take the logarithm of $y$ and treat log $y$ as the response variable. While this will usually bring the assumptions closer to reality, the interpretation of the coefficients is a little more complex. For example, if instead of taking restoration lifespan we had taken $\log_e$ of the life-span, the values of $b_1$, $b_2$, and $b_3$ would indicate the average change in $\log_e y$ for each unit increase in $x_1$, $x_2$, and $x_3$, respectively. Consequently, in order to interpret $b_1$, $b_2$ and $b_3$, it is necessary to take their antilogs, where, for example, antilog $b_1 = e^{b_1}$.

Antilog $b_1$ is a ratio and if, for example, $b_3$ were $0 \cdot 0953$, $e^{0 \cdot 0953} = 1 \cdot 1$ which implies that, adjusting for volume and replacement level, the lifespan is 10% higher for gold than for amalgam restorations. A model in which the response variable has been transformed by the logarithm is called a log–linear or multiplicative model, because the interpretation yields a multiplicative effect; the lifespan for amalgam is *multiplied* by $1 \cdot 1$ to obtain the mean lifespan for gold.

Another complication that very commonly arises is that rather than $y$ being a Normally distributed quantitative variable, the response is the proportion of individuals who have a characteristic. For example, we may be investigating social and economic variables to see how they affect the proportion of adults who have visited their dentist in the last year. For a given combination of explanatory variables, there may be $n$ adults, of whom $r$ visited their dentist in the last year, so that for this subgroup of data, $p$ the proportion is $r/n$, and this is the response variable. Now $p$ is neither Normally distributed nor has constant variance, so it is necessary to transform $p$. A common transformation employed here is $y = \text{logit } p$ where

$$y = \text{logit } p = \log_e[p/(1 - p)]$$

The quantity $p/(1 - p)$ is often called the *odds*. If, for example, 30% of adults visited their dentist in the last year, the odds are 30 to 70, that is, $30/70 = 0 \cdot 43$ and $y = \log_e \text{odds} = \log_e 0 \cdot 43 = -0 \cdot 84$. Regression using $\log_e[p/(1-p)]$ as the response variable is called logistic regression. The calculated coefficients $b_1$, $b_2$, $b_3$, etc, have to be 'antilogged' to give an odds ratio for the independent effect of each explanatory variable. Thus, if $x_3$ defined gender (with $x_3 = 0$ for males and $x_3 = 1$ for females) and $b_3$ turned out to be $0 \cdot 4$, the value of the odds ratio would be $e^{0 \cdot 4} = 1 \cdot 5$. That is, the odds on a woman having visited a dentist is $1 \cdot 5$ times as high as the odds for a man.

Logistic regression is widely used in epidemiology, where the response is the risk of a disease or death. For example, the results of such a study might conclude that in an Asian community, betel-nut chewers have five times the risk of developing oral cancer compared to non-chewers. A word of caution, however. Logistic regression is a special form of multiple regression, and ordinary computer regression programs cannot be used for logistic regression. For a useful review of commonly available programs for multiple regression analysis, readers are referred to Evans.[1]

Thus, the modern medical statistician needs a computer to apply the latest refined methods of statistical analysis, and these techniques may detect effects that might be missed by simpler methods. Alternatively, biases may be removed if the effects of several factors can be disentangled. However, clinically important differences or effects can very often be detected using the simpler univariate methods described in the earlier chapters in this book. Simple methods, coupled with clear presentation and interpretation of results, will often more than adequately justify the conclusions drawn from well planned and executed dental research projects.

## References

1   Evans S J W. Uses and abuses of multivariate methods in epidemiology. *J Epidemiol Community Health* 1988; **42**: 311–315.

# Appendix 1

Critical values of $t$ and $\chi^2$

| df | Critical values of $t$ | | | Critical values of $\chi^2$ | | |
|---|---|---|---|---|---|---|
| | $P=5\%$ | $P=1\%$ | $P=0\cdot1\%$ | $P=5\%$ | $P=1\%$ | $P=0\cdot1\%$ |
| 1 | 12·71 | 63·66 | 636·62 | 3·84 | 6·63 | 10·83 |
| 2 | 4·30 | 9·93 | 31·60 | 5·99 | 9·21 | 13·81 |
| 3 | 3·18 | 5·84 | 12·92 | 7·81 | 11·34 | 16·27 |
| 4 | 2·78 | 4·60 | 8·61 | 9·49 | 13·28 | 18·47 |
| 5 | 2·57 | 4·03 | 6·87 | 11·07 | 15·09 | 20·52 |
| 6 | 2·45 | 3·71 | 5·96 | 12·59 | 16·81 | 22·46 |
| 7 | 2·37 | 3·50 | 5·41 | 14·07 | 18·48 | 24·32 |
| 8 | 2·31 | 3·36 | 5·04 | 15·51 | 20·09 | 26·12 |
| 9 | 2·26 | 3·25 | 4·78 | 16·92 | 21·67 | 27·88 |
| 10 | 2·23 | 3·17 | 4·59 | 18·31 | 23·21 | 29·59 |
| 11 | 2·20 | 3·11 | 4·44 | 19·68 | 24·73 | 31·26 |
| 12 | 2·18 | 3·06 | 4·32 | 21·03 | 26·22 | 32·91 |
| 13 | 2·16 | 3·01 | 4·22 | 22·36 | 27·69 | 34·53 |
| 14 | 2·15 | 2·98 | 4·14 | 23·68 | 29·14 | 36·12 |
| 15 | 2·13 | 2·95 | 4·07 | 25·00 | 30·58 | 37·70 |
| 16 | 2·12 | 2·92 | 4·02 | 26·30 | 32·00 | 39·25 |
| 17 | 2·11 | 2·90 | 3·97 | 27·59 | 33·41 | 40·79 |
| 18 | 2·10 | 2·88 | 3·92 | 28·87 | 34·81 | 42·31 |
| 19 | 2·09 | 2·86 | 3·88 | 30·14 | 36·19 | 43·82 |
| 20 | 2·09 | 2·85 | 3·85 | 31·41 | 37·57 | 45·31 |
| 21 | 2·08 | 2·83 | 3·82 | 32·67 | 38·93 | 46·80 |
| 22 | 2·07 | 2·82 | 3·79 | 33·92 | 40·29 | 48·27 |
| 23 | 2·07 | 2·81 | 3·77 | 35·17 | 41·64 | 49·73 |
| 24 | 2·06 | 2·80 | 3·75 | 36·43 | 42·98 | 51·18 |
| 25 | 2·06 | 2·79 | 3·73 | 37·65 | 44·31 | 52·62 |
| 26 | 2·06 | 2·76 | 3·71 | 38·89 | 45·64 | 54·05 |
| 27 | 2·05 | 2·77 | 3·69 | 40·11 | 46·96 | 55·48 |
| 28 | 2·05 | 2·76 | 3·67 | 41·34 | 48·28 | 56·89 |
| 29 | 2·05 | 2·76 | 3·66 | 42·56 | 49·59 | 58·30 |
| 30 | 2·04 | 2·75 | 3·65 | 43·77 | 50·89 | 59·70 |
| 40 | 2·02 | 2·70 | 3·55 | 55·76 | 63·69 | 73·40 |
| 50 | 2·01 | 2·68 | 3·49 | 67·50 | 76·15 | 86·66 |
| 60 | 2·00 | 2·66 | 3·46 | 79·08 | 88·38 | 99·61 |
| 70 | 1·99 | 2·65 | 3·43 | | | |
| 80 | 1·99 | 2·64 | 3·41 | | | |
| 90 | 1·99 | 2·63 | 3·40 | | | |
| 100 | 1·98 | 2·63 | 3·39 | | | |
| ∞ | 1·96 | 2·58 | 3·29 | | | |

# Appendix 2

## The F Distribution for $P=5\%$

| $df_1 =$ | 1 | 2 | 3 | 4 | 5 | 6 | 7 | 8 | 10 | 15 | 20 | 30 | ∞ |
|---|---|---|---|---|---|---|---|---|---|---|---|---|---|
| $df_2 = 1$ | 161·4 | 199·5 | 215·7 | 224·6 | 230·2 | 234·0 | 236·8 | 238·9 | 241·9 | 245·9 | 248·0 | 250·1 | 254·3 |
| 2 | 18·5 | 19·0 | 19·2 | 19·2 | 19·3 | 19·3 | 19·4 | 19·4 | 19·4 | 19·4 | 19·4 | 19·5 | 19·5 |
| 3 | 10·13 | 9·55 | 9·28 | 9·12 | 9·01 | 8·94 | 8·89 | 8·85 | 8·79 | 8·70 | 8·66 | 8·62 | 8·53 |
| 4 | 7·71 | 6·94 | 6·59 | 6·39 | 6·26 | 6·16 | 6·09 | 6·04 | 5·96 | 5·86 | 5·80 | 5·75 | 5·63 |
| 5 | 6·61 | 5·79 | 5·41 | 5·19 | 5·05 | 4·95 | 4·88 | 4·82 | 4·74 | 4·62 | 4·56 | 4·50 | 4·36 |
| 6 | 5·99 | 5·14 | 4·76 | 4·53 | 4·39 | 4·28 | 4·21 | 4·15 | 4·06 | 3·94 | 3·87 | 3·81 | 3·67 |
| 7 | 5·59 | 4·74 | 4·35 | 4·12 | 3·97 | 3·87 | 3·79 | 3·73 | 3·64 | 3·51 | 3·44 | 3·38 | 3·23 |
| 8 | 5·32 | 4·46 | 4·07 | 3·84 | 3·69 | 3·58 | 3·50 | 3·44 | 3·35 | 3·22 | 3·15 | 3·08 | 2·93 |
| 9 | 5·12 | 4·26 | 3·86 | 3·63 | 3·48 | 3·37 | 3·29 | 3·23 | 3·14 | 3·01 | 2·94 | 2·86 | 2·71 |
| 10 | 4·96 | 4·10 | 3·71 | 3·48 | 3·33 | 3·22 | 3·14 | 3·07 | 2·98 | 2·85 | 2·77 | 2·70 | 2·54 |
| 11 | 4·84 | 3·98 | 3·59 | 3·36 | 3·20 | 3·09 | 3·01 | 2·95 | 2·85 | 2·72 | 2·65 | 2·57 | 2·40 |
| 12 | 4·75 | 3·89 | 3·49 | 3·26 | 3·11 | 3·00 | 2·91 | 2·85 | 2·75 | 2·62 | 2·54 | 2·47 | 2·30 |
| 13 | 4·67 | 3·81 | 3·41 | 3·18 | 3·03 | 2·92 | 2·83 | 2·77 | 2·67 | 2·53 | 2·46 | 2·38 | 2·21 |
| 14 | 4·60 | 3·74 | 3·34 | 3·11 | 2·96 | 2·85 | 2·76 | 2·70 | 2·60 | 2·46 | 2·39 | 2·31 | 2·13 |
| 15 | 4·54 | 3·68 | 3·29 | 3·06 | 2·90 | 2·79 | 2·71 | 2·64 | 2·54 | 2·40 | 2·33 | 2·25 | 2·07 |
| 16 | 4·49 | 3·63 | 3·24 | 3·01 | 2·85 | 2·74 | 2·66 | 2·59 | 2·49 | 2·35 | 2·28 | 2·19 | 2·01 |
| 17 | 4·45 | 3·59 | 3·20 | 2·96 | 2·81 | 2·70 | 2·61 | 2·55 | 2·45 | 2·31 | 2·23 | 2·15 | 1·96 |
| 18 | 4·41 | 3·55 | 3·16 | 2·93 | 2·77 | 2·66 | 2·58 | 2·51 | 2·41 | 2·27 | 2·19 | 2·11 | 1·92 |
| 19 | 4·38 | 3·52 | 3·13 | 2·90 | 2·74 | 2·63 | 2·54 | 2·48 | 2·38 | 2·23 | 2·16 | 2·07 | 1·88 |
| 20 | 4·35 | 3·49 | 3·10 | 2·87 | 2·71 | 2·60 | 2·51 | 2·45 | 2·35 | 2·20 | 2·12 | 2·04 | 1·84 |
| 21 | 4·32 | 3·47 | 3·07 | 2·84 | 2·68 | 2·57 | 2·49 | 2·42 | 2·32 | 2·18 | 2·10 | 2·01 | 1·81 |
| 22 | 4·30 | 3·44 | 3·05 | 2·82 | 2·66 | 2·55 | 2·46 | 2·40 | 2·30 | 2·15 | 2·07 | 1·98 | 1·78 |
| 23 | 4·28 | 3·42 | 3·03 | 2·80 | 2·64 | 2·53 | 2·44 | 2·37 | 2·27 | 2·13 | 2·05 | 1·96 | 1·76 |
| 24 | 4·26 | 3·40 | 3·01 | 2·78 | 2·62 | 2·51 | 2·42 | 2·36 | 2·25 | 2·11 | 2·03 | 1·94 | 1·73 |
| 25 | 4·24 | 3·39 | 2·99 | 2·76 | 2·60 | 2·49 | 2·40 | 2·34 | 2·24 | 2·09 | 2·01 | 1·92 | 1·71 |
| 26 | 4·23 | 3·37 | 2·98 | 2·74 | 2·59 | 2·47 | 2·39 | 2·32 | 2·22 | 2·07 | 1·99 | 1·90 | 1·69 |
| 27 | 4·21 | 3·35 | 2·96 | 2·73 | 2·57 | 2·46 | 2·37 | 2·31 | 2·20 | 2·06 | 1·97 | 1·88 | 1·67 |
| 28 | 4·20 | 3·34 | 2·95 | 2·71 | 2·56 | 2·45 | 2·36 | 2·29 | 2·19 | 2·04 | 1·96 | 1·87 | 1·65 |
| 29 | 4·18 | 3·33 | 2·93 | 2·70 | 2·55 | 2·43 | 2·35 | 2·28 | 2·18 | 2·03 | 1·94 | 1·85 | 1·64 |
| 30 | 4·17 | 3·32 | 2·92 | 2·69 | 2·53 | 2·42 | 2·33 | 2·27 | 2·16 | 2·01 | 1·93 | 1·84 | 1·62 |
| 40 | 4·08 | 3·23 | 2·84 | 2·61 | 2·45 | 2·34 | 2·25 | 2·18 | 2·08 | 1·92 | 1·84 | 1·74 | 1·51 |
| 60 | 4·00 | 3·15 | 2·76 | 2·53 | 2·37 | 2·25 | 2·17 | 2·10 | 1·99 | 1·84 | 1·75 | 1·65 | 1·39 |
| 120 | 3·92 | 3·07 | 2·68 | 2·45 | 2·29 | 2·17 | 2·09 | 2·02 | 1·91 | 1·75 | 1·66 | 1·55 | 1·25 |
| ∞ | 3·84 | 3·00 | 2·60 | 2·37 | 2·21 | 2·10 | 2·01 | 1·94 | 1·83 | 1·67 | 1·57 | 1·46 | 1·00 |

# Appendix 2 (continued)

## The F Distribution for P = 1%

| df₁ = | 1 | 2 | 3 | 4 | 5 | 6 | 7 | 8 | 10 | 15 | 20 | 30 | ∞ |
|---|---|---|---|---|---|---|---|---|---|---|---|---|---|
| df₂ = 1 | 4052 | 5000 | 5403 | 5625 | 5764 | 5859 | 5928 | 5981 | 6056 | 6157 | 6209 | 6261 | 6366 |
| 2 | 98·5 | 99·0 | 99·2 | 99·2 | 99·3 | 99·3 | 99·4 | 99·4 | 99·4 | 99·4 | 99·4 | 99·5 | 99·5 |
| 3 | 34·12 | 30·82 | 29·46 | 28·71 | 28·24 | 27·91 | 27·67 | 27·49 | 27·23 | 26·87 | 26·69 | 26·50 | 26·13 |
| 4 | 21·20 | 18·00 | 16·69 | 15·98 | 15·52 | 15·21 | 14·98 | 14·80 | 14·55 | 14·20 | 14·02 | 13·84 | 13·46 |
| 5 | 16·26 | 13·27 | 12·06 | 11·39 | 10·97 | 10·67 | 10·46 | 10·29 | 10·05 | 9·72 | 9·55 | 9·38 | 9·02 |
| 6 | 13·75 | 10·92 | 9·78 | 9·15 | 8·75 | 8·47 | 8·26 | 8·10 | 7·87 | 7·56 | 7·40 | 7·23 | 6·88 |
| 7 | 12·25 | 9·55 | 8·45 | 7·85 | 7·46 | 7·19 | 6·99 | 6·84 | 6·62 | 6·31 | 6·16 | 5·99 | 5·65 |
| 8 | 11·26 | 8·65 | 7·59 | 7·01 | 6·63 | 6·37 | 6·18 | 6·03 | 5·81 | 5·52 | 5·36 | 5·20 | 4·86 |
| 9 | 10·56 | 8·02 | 6·99 | 6·42 | 6·06 | 5·80 | 5·61 | 5·47 | 5·26 | 4·96 | 4·81 | 4·65 | 4·31 |
| 10 | 10·04 | 7·56 | 6·55 | 5·99 | 5·64 | 5·39 | 5·20 | 5·06 | 4·85 | 4·56 | 4·41 | 4·25 | 3·91 |
| 11 | 9·65 | 7·21 | 6·22 | 5·67 | 5·32 | 5·07 | 4·89 | 4·74 | 4·54 | 4·25 | 4·10 | 3·94 | 3·60 |
| 12 | 9·33 | 6·93 | 5·95 | 5·41 | 5·06 | 4·82 | 4·64 | 4·50 | 4·30 | 4·01 | 3·86 | 3·70 | 3·36 |
| 13 | 9·07 | 6·70 | 5·74 | 5·21 | 4·86 | 4·62 | 4·44 | 4·30 | 4·10 | 3·82 | 3·66 | 3·51 | 3·17 |
| 14 | 8·86 | 6·51 | 5·56 | 5·04 | 4·69 | 4·46 | 4·28 | 4·14 | 3·94 | 3·66 | 3·51 | 3·35 | 3·00 |
| 15 | 8·68 | 6·36 | 5·42 | 4·89 | 4·56 | 4·32 | 4·14 | 4·00 | 3·80 | 3·52 | 3·37 | 3·21 | 2·87 |
| 16 | 8·53 | 6·23 | 5·29 | 4·77 | 4·44 | 4·20 | 4·03 | 3·89 | 3·69 | 3·41 | 3·26 | 3·10 | 2·75 |
| 17 | 8·40 | 6·11 | 5·18 | 4·67 | 4·34 | 4·10 | 3·93 | 3·79 | 3·59 | 3·31 | 3·16 | 3·00 | 2·65 |
| 18 | 8·29 | 6·01 | 5·09 | 4·58 | 4·25 | 4·01 | 3·84 | 3·71 | 3·51 | 3·23 | 3·08 | 2·92 | 2·57 |
| 19 | 8·18 | 5·93 | 5·01 | 4·50 | 4·17 | 3·94 | 3·77 | 3·63 | 3·43 | 3·15 | 3·00 | 2·84 | 2·49 |
| 20 | 8·10 | 5·85 | 4·94 | 4·43 | 4·10 | 3·87 | 3·70 | 3·56 | 3·37 | 3·09 | 2·94 | 2·78 | 2·42 |
| 21 | 8·02 | 5·78 | 4·87 | 4·37 | 4·04 | 3·81 | 3·64 | 3·51 | 3·31 | 3·03 | 2·88 | 2·72 | 2·36 |
| 22 | 7·95 | 5·72 | 4·82 | 4·31 | 3·99 | 3·76 | 3·59 | 3·45 | 3·26 | 2·98 | 2·83 | 2·67 | 2·31 |
| 23 | 7·88 | 5·66 | 4·76 | 4·26 | 3·94 | 3·71 | 3·54 | 3·41 | 3·21 | 2·93 | 2·78 | 2·62 | 2·26 |
| 24 | 7·82 | 5·61 | 4·72 | 4·22 | 3·90 | 3·67 | 3·50 | 3·36 | 3·17 | 2·89 | 2·74 | 2·58 | 2·21 |
| 25 | 7·77 | 5·57 | 4·68 | 4·18 | 3·85 | 3·63 | 3·46 | 3·32 | 3·13 | 2·85 | 2·70 | 2·54 | 2·17 |
| 26 | 7·72 | 5·53 | 4·64 | 4·14 | 3·82 | 3·59 | 3·42 | 3·29 | 3·09 | 2·81 | 2·66 | 2·50 | 2·13 |
| 27 | 7·68 | 5·49 | 4·60 | 4·11 | 3·78 | 3·56 | 3·39 | 3·26 | 3·06 | 2·78 | 2·63 | 2·47 | 2·10 |
| 28 | 7·64 | 5·45 | 4·57 | 4·07 | 3·75 | 3·53 | 3·36 | 3·23 | 3·03 | 2·75 | 2·60 | 2·44 | 2·06 |
| 29 | 7·60 | 5·42 | 4·54 | 4·04 | 3·73 | 3·50 | 3·33 | 3·20 | 3·00 | 2·73 | 2·57 | 2·41 | 2·03 |
| 30 | 7·56 | 5·39 | 4·51 | 4·02 | 3·70 | 3·47 | 3·30 | 3·17 | 2·98 | 2·70 | 2·55 | 2·39 | 2·01 |
| 40 | 7·31 | 5·18 | 4·31 | 3·83 | 3·51 | 3·29 | 3·12 | 2·99 | 2·80 | 2·52 | 2·37 | 2·20 | 1·80 |
| 60 | 7·08 | 4·98 | 4·13 | 3·65 | 3·34 | 3·12 | 2·95 | 2·82 | 2·63 | 2·35 | 2·20 | 2·03 | 1·60 |
| 120 | 6·85 | 4·79 | 3·95 | 3·48 | 3·17 | 2·96 | 2·79 | 2·66 | 2·47 | 2·19 | 2·03 | 1·86 | 1·38 |
| ∞ | 6·63 | 4·61 | 3·78 | 3·32 | 3·02 | 2·80 | 2·64 | 2·51 | 2·32 | 2·04 | 1·88 | 1·70 | 1·00 |

# Appendix 3

## Solutions to practical examples

### Chapter 2:   Simple summary calculations

For the sample *mean:*

$$\bar{x} = \Sigma x/n = ((1\times3)+(2\times4)+(3\times6)+(4\times3)+(5\times3)+(6\times1))/20 = 3\cdot1 \text{ kg}$$

If all the observations are arrayed in ascending or descending order of magnitude, the *median* will be the average of the middle two observations, in this case, the 10th and 11th. In ascending order we have

$$1 \quad 1 \quad 1 \quad 2 \quad 2 \quad 2 \quad 2 \quad 3 \quad 3 \quad 3 \quad 3 \quad 3 \quad 3 \ldots$$

At this point we need go no higher, because we have already identified the 10th and 11th observations; they are both 3. Therefore the median value is 3 kg.

The modal observation is the one that occurs most frequently. In this case it is 3, which occurs six times. The *mode* is therefore 3 kg.

The easiest way to calculate the standard deviation is by calculator or computer. If, however, you are doing it from first principles, then the figures you need are

$$\Sigma x = 62 \quad \Sigma x^2 = 232 \quad (\Sigma x)^2/n = 192\cdot2 \quad \Sigma(x-\bar{x})^2 = 232-192\cdot2 = 39\cdot8$$
$$s^2 = 39\cdot8/(20-1) = 2\cdot095 \quad s = \sqrt{2\cdot095} = 1\cdot45.$$

This means that in a large sample we would expect roughly 95% of the breakages to occur within the range:

$$\bar{x}\pm2\times1\cdot45 = \bar{x}\pm2\cdot90 = 0\cdot2 \text{ kg to } 6\cdot0 \text{ kg.}$$

Such a wide range of breaking points, coupled with the possibility of quite low breaking stresses, suggest that the alloy is quite unsuitable for use in the mouth.

### Chapter 3: Probability and sampling

The sample mean ($\bar{x}$) is $0\cdot40$ ppm. The standard deviation is $0\cdot03$ ppm. The standard error of the mean is therefore $0\cdot03/\sqrt{12} = 0\cdot00866$.

Ninety-five per cent confidence interval for the population mean is therefore

$$0\cdot4\pm1\cdot96\times0\cdot00866 = 0\cdot40\pm0\cdot01697 = 0\cdot38 \text{ ppm to } 0\cdot42 \text{ ppm.}$$

It is therefore unlikely that the mean level of fluoride in this region is less than $0\cdot38$ ppm or higher than $0\cdot42$ ppm. The levels of fluoride throughout the region are likely to range between $0\cdot34$ and $0\cdot46$ ($0\cdot40\pm1\cdot96\sigma$), in other words, reasonably constant.

## Chapter 4: Significance tests

The estimated standard error of the mean is $30 \cdot 3/\sqrt{10} = 9 \cdot 582$

$$t = (380 - 356)/9 \cdot 582 = 24/9 \cdot 582 = 2 \cdot 50 \text{ on 9df.}$$

The 5% critical value of $t$ with 9df is $2 \cdot 26$. Since $2 \cdot 50$ is greater than this critical value, it follows that there is less than a 5% chance that the difference between the two means is just the result of sampling error.

The 95% confidence interval for the true difference is

$$24 \pm 2 \cdot 26 \times 9 \cdot 582 = 24 \pm 21 \cdot 655 = 2 \cdot 3 \text{ hours to } 45 \cdot 6 \text{ hours}$$

In other words, we would expect the new lamps to last, on average, between $2 \cdot 3$ hours and $45 \cdot 6$ hours longer than the old ones. If the new lamps cost only 5% more than the old, it is certainly worth investing in the new ones. However, if the price difference is, say, 10% more, then the difference in lifespans, although statistically significant, would not be economically significant!

## Chapter 5: Significance tests continued

| | A | B |
|---|---|---|
| $n$ | 14 | 14 |
| $\Sigma x$ | 153 | 129 |
| $\Sigma x^2$ | 1739 | 1253 |
| $\bar{x}$ | $10 \cdot 93$ | $9 \cdot 21$ |
| $(\Sigma x)^2/n$ | $1672 \cdot 071$ | $1188 \cdot 643$ |
| $\Sigma(x - \bar{x})^2$ | $66 \cdot 929$ | $64 \cdot 357$ |
| $s^2$ | $5 \cdot 148$ | $4 \cdot 951$ |
| $s$ | $2 \cdot 269$ | $2 \cdot 225$ |

$$\text{pooled estimate } s^2 = (66 \cdot 929 + 64 \cdot 357)/26 = 5 \cdot 0495$$
$$s = \sqrt{5 \cdot 0495} = 2 \cdot 247$$
$$\text{Est. } SE(\bar{x}_A - \bar{x}_B) = 2 \cdot 247\sqrt{(1/14 + 1/14)} = 0 \cdot 8493$$
$$t = (10 \cdot 93 - 9 \cdot 21)/0 \cdot 8493 = 2 \cdot 02 \text{ on 26df.}$$

This $t$ value is not statistically significant at the 5% level. We therefore have insufficient evidence to suggest that the difference in mean number of patients between the two hospital departments is due to anything other than sampling error.

A 95% confidence interval for the true difference:

$$1 \cdot 72 \pm 2 \cdot 06 \times 0 \cdot 8493$$
$$= 1 \cdot 72 \pm 1 \cdot 750$$
$$= -0 \cdot 03 \text{ to } 3 \cdot 47$$

This confirms that from these figures we could expect an average of anything between $0 \cdot 03$ patients more per session in Hospital B to $3 \cdot 5$ patients more per session in Hospital A.

## Chapter 6: Comparison of proportions. Chi-squared.

### (1) *Chi-squared*

Observed values:

|             | Male | Female | Total |
|-------------|------|--------|-------|
| Had surgery | 47   | 80     | 127   |
| No surgery  | 145  | 404    | 549   |
| Total       | 192  | 484    | 676   |

Expected values:

|             | Male    | Female  |
|-------------|---------|---------|
| Had surgery | 36·07   | 90·93   |
| No surgery  | 155·93  | 393·07  |

Chi-squared:

$(47-36\cdot07)^2/36\cdot07]+(80-90\cdot93)^2/90\cdot93]$
$+(145-155\cdot93)^2/155\cdot93]+(404-393\cdot07)^2/393\cdot07]$
$= 3\cdot3120+1\cdot3138+0\cdot7661+0\cdot3039$
$= 5\cdot6958$ on 1df.

The 5% critical value of $\chi^2$ on 1 df is 3·84; there is therefore a significant difference between the proportions of men and women undergoing surgery.

### (2) *SND test of proportions*

|             | Male    | Female  |
|-------------|---------|---------|
| $n$         | 192     | 484     |
| Had surgery | 47      | 80      |
| Proportion  | 0·2448  | 0·1653  |

Pooled estimate of $\pi = (47+80)/(192+484) = 0\cdot1879$

$\mathrm{SE}(p_M-p_F) = \sqrt{0\cdot1879(1-0\cdot1879)(1/192+1/484)}$
$= 0\cdot03332$

SND $= (0\cdot2448-0\cdot1653)/0\cdot03332 = 0\cdot0795/0\cdot03332 = 2\cdot39$

Since this exceeds 1·96, the critical value of the SND at the 5% level of significance, the result is statistically significant. However, 2·39 is not greater than the critical SND for the 1% level, which is 2·58. The significance level is therefore $0\cdot01<P<0\cdot05$.

The 95% confidence interval for the true difference between the proportions for men and women is

$$0\cdot0795\pm1\cdot96\times0\cdot03332$$
$$= 0\cdot0795\pm0\cdot06531$$
$$= 0\cdot0142 \text{ to } 0\cdot1448$$

In other words, the difference between the two proportions is unlikely to be less than 1·4% or greater than 14·5%.

## Chapter 7: Linear regression and correlation

$\bar{x}$ = 22·1111   $\bar{y}$ = 52   $n$ = 9
$\Sigma x$ = 199   $\Sigma x^2$ = 4575
$\Sigma y$ = 468   $\Sigma y^2$ = 24928   $\Sigma xy$ = 10640
X = 4575 − (199²/9) = 174·8889
Y = 24928 − (468²/9) = 592
Z = 10640 − (199×468/9) = 292

$a$ = 52 − 22·1111(292/174·8889) = 15·0826
$b$ = 292/174·8889 = 1·6696

$y$ = 15·0826 + 1·6696$x$

If $x$ = 20, $y$ = 15·0826 + (1·6696×20) = 48·475

Var(y) = [(592 − (292²/174·8889))/(9 − 2)][1 + 1/9 + ((20 − 22·1111)²/174·8889)]
= 16·9625
SE(y) = $\sqrt{16.9625}$ = 4·11855

95% Conference Interval for $y$:

48·475 ± 2·37 × 4·11855
= 48·475 ± 9·7610
= 38·7 to 58·2

In other words, a student scoring 20 in the 'rehearsal' is likely to score between 39 and 58 in the real examination. This suggests that the rehearsal is not a particularly accurate predictor of the score in the real examination.

Correlation coefficient $r$ = 292/$\sqrt{(174·8889 × 592)}$ = 0·907.
Variation of $y$ explained by $x$ = 0·907² = 0·8235 or 82%.

## Chapter 8: Comparing several sample means

|  | Group | | | |  |
|---|---|---|---|---|---|
|  | 1 | 2 | 3 | 4 |  |
|  | 8·0 | 8·3 | 8·5 | 7·3 |  |
|  | 7·5 | 6·8 | 8·3 | 7·2 |  |
|  | 8·2 | 7·2 | 7·9 | 6·8 |  |
|  | 7·5 | 6·7 | 8·2 | 6·7 |  |
|  | 7·3 |  | 8·4 |  |  |

| | 1 | 2 | 3 | 4 | |
|---|---|---|---|---|---|
| $n$ | 5 | 4 | 5 | 4 | 18 |
| $\bar{x}$ | 7·7 | 7·25 | 8·26 | 7·0 | |
| $\Sigma x$ | 38·5 | 29·0 | 41·3 | 28·0 | 136·8 |
| $\Sigma x^2$ | 297·03 | 211·86 | 341·35 | 196·26 | |
| $\Sigma(x-\bar{x})^2$ | 0·58 | 1·61 | 0·212 | 0·26 | |
| $s$ | 0·3808 | 0·7326 | 0·2302 | 0·2944 | |

$$s^2 = (0·58 + 1·61 + 0·212 + 0·26)/(4+3+4+3) = 0·1901$$

Between Group SSq $= (38·5^2/5 + 29·0^2/4 + 41·3^2/5 + 28·0^2/4 - 136·8^2/18)/3 = 4·158$
Msq $= 4·158/3 = 1·386$

Total SSq $= 8·0^2 + 8·3^2 + 8·5^2 + 7·3^2 + 7·5^2 + \ldots - 136·8^2/18 = 6·82$

Analysis of variance:

|  | df | SSQ | MSQ | F |
|---|---|---|---|---|
| Between groups | 3 | 4·158 | 1·386 | 7·295 |
| Within groups | 14 | 2·662 | 0·1901 | |
| Total | 17 | 6·8200 | | |

(These results would nowadays of course be obtained with the aid of a computer.) There is therefore evidence to suggest that mean incisor width is indeed associated with ethnic grouping.

|  | 1 | 2 | 3 | 4 |
|---|---|---|---|---|
| Mean (mm) | 7·7 | 7·25 | 8·26 | 7·0 |
| SE($\bar{x}$) | 0·195 | 0·218 | 0·195 | 0·218 |
| 95% CI | 7·16 | 6·56 | 7·72 | 6·31 |
|  | – | – | – | – |
|  | 8·24 | 7·94 | 8·80 | 7·69 |

Note: These standard errors are calculated using the pooled variance of all the samples (ie the within-group mean square). Thus $s = \sqrt{0·1901} = 0·4360$, and for Group 1, SE($\bar{x}$) $= 0·4360/\sqrt{5} = 0·1950$.

## Chapter 9: Examiner variability

|  |  | Exam I | | |
|---|---|---|---|---|
|  |  | Sound | Carious | Total |
| Exam II | Sound | 130(0·65) | 10(0·05) | 140(0·70) |
|  | Carious | 6(0·03) | 54(0·27) | 60(0·30) |
|  | Total | 136(0·68) | 64(0·32) | 200 |

Dice (sound) $= 0·65/((0·68+0·70)/2) = 0·94$ or 94%
Dice (carious) $= 0·27/((0·32+0·30)/2) = 0·87$ or 87%

Thus, there is a 94% probability of repeating a 'sound' diagnosis, and an 87% probability of repeating a 'carious' diagnosis.
  In the survey:

500 children represent the equivalent of $500 \times 20 = 10000$ teeth.
Of these 10000, $0·15 \times 500 = 74$ are missing, leaving 9925.
Of these 9925, $0·24 \times 500 = 120$ have been filled, leaving 9805.
Of these 9805, $0·95 \times 500 = 475$ were marked carious, leaving 9330.
At the initial survey, therefore, 9330 teeth were recorded as 'sound', and 475 were recorded as 'carious'.

  Applying the 94% 'sound' Dice figure to the total 'sound':
Of 9330 teeth marked 'sound' initially, at least $9330 \times 0·94 = 8770·2$ would be marked 'sound' on the second occasion, but up to $9330(1-0·94) = 559·8$ could now be marked 'carious'.
Similarly, of 475 teeth marked 'carious' initially, at least $475 \times 0·87 = 413·25$ would be marked 'carious' on the second occasion, but up to $475(1-0·87) = 61·75$ could now be marked 'sound'.
Thus, the total which could be marked as 'sound' on the second occasion would be $8770·2 + 61·75 = 8832$ (rounding to a whole number).
Similarly, the total which could be marked as 'carious' on the second occasion would be $413·25 + 559·8 = 973$.
Thus, the new dt value could be as high as $973/500 = 1·95$ and the new dmft would then be $1·95 + 0·15 + 0·24 = 2·34$ instead of the original 1·33, a 76% increase! (This is actually misleading, because the percentage is inflated due to the small numbers involved.)

# Index